AF896669

Minimum Quantity Lubrication Machining

Xia Ji

Minimum Quantity Lubrication Machining

Process Analysis and Analytical Modeling

Springer

Xia Ji
School of Mechanical Engineering
Donghua University
Shanghai, China

ISBN 978-981-19-7086-3 ISBN 978-981-19-7087-0 (eBook)
https://doi.org/10.1007/978-981-19-7087-0

© The Editor(s) (if applicable) and The Author(s), under exclusive license to Springer Nature Singapore Pte Ltd. 2023
This work is subject to copyright. All rights are solely and exclusively licensed by the Publisher, whether the whole or part of the material is concerned, specifically the rights of translation, reprinting, reuse of illustrations, recitation, broadcasting, reproduction on microfilms or in any other physical way, and transmission or information storage and retrieval, electronic adaptation, computer software, or by similar or dissimilar methodology now known or hereafter developed.
The use of general descriptive names, registered names, trademarks, service marks, etc. in this publication does not imply, even in the absence of a specific statement, that such names are exempt from the relevant protective laws and regulations and therefore free for general use.
The publisher, the authors, and the editors are safe to assume that the advice and information in this book are believed to be true and accurate at the date of publication. Neither the publisher nor the authors or the editors give a warranty, expressed or implied, with respect to the material contained herein or for any errors or omissions that may have been made. The publisher remains neutral with regard to jurisdictional claims in published maps and institutional affiliations.

This Springer imprint is published by the registered company Springer Nature Singapore Pte Ltd.
The registered company address is: 152 Beach Road, #21-01/04 Gateway East, Singapore 189721, Singapore

Foreword

Minimum quantity lubrication (MQL) in machining is a technology based on the concept of near-dry machining to minimize the use of cutting fluid while not compromising the cooling and lubricating benefits of the cutting fluid. This technology has gained much acceptance in machining due to its advantages of retaining mechanical and thermal effectiveness, reducing coolant cost, easing part cleaning, and possibly increasing environmental-friendliness. It has been found that, under certain cutting conditions, MQL machining bears the potential merits of offering comparable or even better machinability compared to traditional dry and wet machining by careful selection of the appropriate machining process and fluid application parameters.

Despite the engineering success in the implementation of MQL, the fundamental understanding of its physics and mechanics is still at an early stage of hypothesizing, particularly on the lubrication and cooling mechanisms of MQL and the thermodynamic behaviors of the part and the fluid during MQL machining. Unfortunately, possible achievement of cutting performance of MQL usually relies upon operator's work experience or excessive trial-and-error empiricism at the current state of technology. A scientific, systematic, and in-depth understanding of MQL process physics and mechanics is essential to create a clean, sustainable, and high-efficient production environment with the maximization of the MQL benefits.

To address these issues, Dr. Ji has spent more than one decade working on MQL process analysis and analytical modeling. The underlying mechanisms of MQL in the machining process has been analyzed and the lubricating and cooling effects on thermodynamic behavior are quantitatively evaluated in detail. The analytical prediction model of MQL is established flanked by fundamental physics and thermodynamic, without relying upon time-demanding experimentations and numerical simulations. It provides a solid platform to support further process planning in achieving specified machined part properties, integrity, and functionalities in a first-and-every-cut-correct manner. This book can serve not only as an ideal textbook for graduate student and undergraduate students of senior grade, but also as a very valuable reference for researchers and engineering practitioners in the field.

I believe that the research results presented in this book will contribute significantly to the development of MQL methodology, as well as the other machining process analysis and modeling from the theoretical, physical, and practical perspectives.

Steven Y. Liang, Ph.D.
Morris M. Bryan, Jr. Professor for
Advanced Manufacturing Systems,
George W. Woodruff School
of Mechanical Engineering
Georgia Institute of Technology
Atlanta, GA, USA

Preface

The machined residual stress seriously affects the dimensional accuracy, dynamic/static strength, fatigue strength, and corrosion resistance of the components, which mainly determine the service life of the products. Therefore, it is very crucial to understand the generation mechanism of the residual stress and accurately predict the machined residual stress for the quality control of the machined components.

Cutting fluid is an important component of the traditional machining process due to its effect on cooling, lubrication, chip removal, rust prevention, etc. It is reported that the cost of cutting fluid is about 7%~17% of the total machining cost. In addition, the implementation of the cutting fluid will also result in environmental pollution and health hazards. Under this background, the minimum quantity lubrication (MQL) machining as a green and sustainable machining technology has attracted more and more attention in industry and academia.

This book focuses on the process analysis and analytical modeling in MQL machining. The cutting force, cutting temperature, and residual stress prediction model in MQL machining are established and verified by the experiments. Based on validation models, the sensitivity analysis of MQL is discussed.

This book provides readers a fast and accurate way to predict the cutting force, cutting temperature, and residual stress in MQL machining, and will give support for further MQL process planning.

I would like to express my sincere appreciation and thankfulness to my doctoral supervisor, Prof. Steven Y. Liang and Prof. Xueping Zhang. This book cannot be completed without their help. The contents covered in this book are supported by National Natural Science Foundation of China (No:52175384, No:51705073).

Shanghai, China Xia Ji

Contents

About the Editor

Ms. Xia Ji received the B.S. degree from Department of Mechanical Engineering, Yancheng Institute of Technology, Yancheng, China in 2005; the M.S. degree from Department of Mechanical Engineering, Jiangsu University, Zhenjiang, China in 2008; and the Ph.D. degree from Shanghai Jiaotong University, Shanghai, China in 2014, respectively. She has worked in the lab of Prof. Steven Y. Liang in Georgia Institute of Technology for one and a half year. She joined Donghua University in 2014 as an Assistant Professor and has been an Associate Professor since 2017. Her research interests focus on the modeling and simulation of thermo-dynamical behavior in metal cutting process.

Chapter 1
Introduction

1.1 Background

With the rapid development of the manufacturing technology, the quality requirements of the machined components are getting higher and higher. The machined components not only require high geometric accuracy, but also superior mechanical properties such as fatigue strength, corrosion resistance, wear resistance and etc. These properties are all closely related to the surface integrity of the machined components. The concept of surface integrity is first introduced by Field and Kahles [1]. It is defined as the inherent or enhanced condition of a surface produced by machining process or other surface generation operations. Machined surface integrity includes many characteristics, such as surface roughness, surface ripple, macroscopic crack, microstructure, residual stress, etc., among which the residual stress is particularly important.

The residual stress distribution is an important index to evaluate the quality of the machined surface, which seriously affects the static strength, fatigue strength, and corrosion resistance of the workpiece. Thus, it determines the service life of the workpiece. At the same time, the residual stress is also the main factor affecting the geometrical stability of the workpiece. The residual stress on the surface/subsurface can increase or decrease the capacity of the workpiece to bear the load in the process of using, which mainly depends on the state, magnitude and distribution of the residual stress on the machined surface.

The residual stress profile at the surface and sub-surface of the machined components can be influenced by many factors, such as cutting parameters, tool geometry, material properties, lubrication conditions, and etc. In addition, different from other characteristics, the machined-induced residual stress cannot be duplicated. Because of this, understanding the residual stress imparted by machining is an important aspect of understanding machining and overall part quality.

© The Author(s), under exclusive license to Springer Nature Singapore Pte Ltd. 2023
X. Ji, *Minimum Quantity Lubrication Machining*,
https://doi.org/10.1007/978-981-19-7087-0_1

At present, the research methods of residual stress mainly include experimental measurement [2, 3], analytical modeling [4], and finite element method [5, 6]. Experimental measurement serves as the ultimate validation tool for numerical or analytical simulations. However, it is time-consuming and laborious, and its research is limited by experimental conditions, and it cannot be tested under extreme conditions. With the rapid development of computer technology, finite element method is more popular in the analysis of machining residual stress. The finite element method can accurately simulate the metal cutting process, so as to effectively predict the cutting temperature, cutting force, and residual stress distribution on the machined surface. However, finite element method needs to recalculate when changing the cutting conditions, which takes a lot of time, and it also cannot do the reverse calculation. Therefore, finite element method is not suitable for process optimization and control. Compared to the experimental measurement and finite element method, analytical modeling method which is based on physical model, provides a fast and accurate method to predict the machining residual stress.

Most of the existing methods for residual stress are tailored for dry machining. However, many machining operations require cutting fluid in order to produce the desired product. Therefore, from a practical aspect, there is a need to predict the effect of cutting fluids on residual stress formation in machining.

In metal cutting process, the use of cutting fluids is the most common strategy to improve the tool life, the product surface finish and the size accuracy. Cutting fluids also make chip-breaking and chip-transport easier. However, the introduction of cutting fluids often produces airborne mist, smoke and other particulates in the shop floor air quality. These products bring the environmental, health and safety concerns. In response to these health effects through skin contact or inhalation of cutting fluid mist, the National Institute for Occupational Safety and Health (NIOSH) has recommended that the permissible exposure level (PEL) is 0.5 mg/m^3 as the metalworking fluid concentration on the shop floor [7, 8]. In addition, the cost of using cutting fluids is several times higher than tool costs. A recent survey indicated that the cost of cutting fluids and the auxiliary equipment compromise nearly 7~17% of the total machining costs. Compared to the cost of the cutting tools (2~4%), the cutting fluid cost is significantly high [9, 10]. As a result, there is a need to reduce the use of the cutting fluids.

The economic and environmental concerns on the use of cutting fluids lead to the research of minimum quantity lubrication (MQL) machining several years ago. MQL machining refers to the use of a small amount of cutting fluid, typically in the order of 100 ml/h or less, which is about ten-thousandth of the amount of cutting fluid used in flood cooled machining process. The concept of MQL machining is based on the principle of loss lubrication with a completely dry surface after the machining process. Therefore, the minimum quantity lubrication machining is also recognized as near dry machining (NDM).

The challenges presented by introducing coolant and lubrication into residual stress include altering the contact stresses due to changes in friction as well as changes in the thermal stress due to cooling. Many scholars have carried out a lot of researches on the residual stress in MQL machining, which are mainly focusing on the experimental measurement or finite element method. They were devoted to study the relationship between the process parameters and the machined residual stress of the workpiece, but few studies were on analytical modeling [11–13]. In view of the limitations of the experimental measurement and the finite element method, it is necessary to analyze the residual stress in MQL machining by analytical modeling method.

This book research launches under this background. Because of the importance of conservation and environmental awareness, the implementation of minimum quantity lubrication machining as a viable alternative to traditional flood cooling is an important research area. It is being paid more and more attention on this topic. Residual stress modeling in minimum quantity lubrication machining is a natural extension of the current state of modeling minimum quantity lubrication machining.

The goal of this research is to develop an analytical model for predicting residual stresses in MQL machining, and to find out how residual stress vary with lubrication conditions. The results of this study could support further planning and optimization of machining process configuration and parameters in achieving desirable residual stress profiles, and it will provide a solid theoretical foundation for the future study.

1.2 Literature Review of Machining Induced Residual Stress

Metal cutting process is a large plastic deformation process accompanied by high temperature, high pressure, and high strain rate. At the same time, due to the effect of cutting force, cutting heat, and the friction between the tool and workpiece, residual stress distribution is formed on the machined surface. The machined residual stress can be divided into tensile residual stress and compressive residual stress. Machining induced residual stress has been a topic of research since the 1950's. In the literature, the determination of residual stresses can be broadly clustered as experimental measurements, finite element calculations, analytical models, and various combinations of these efforts. In the following, the state of art of the residual stress is summarized from these three aspects.

1.2.1 Experimental Efforts in Residual Stress

Most of the early efforts at determining the effect of machining on residual stress were experiments in nature. Since the pioneering works in 1950s, a substantial amount of experimental work has accumulated regarding the development of residual stresses as a function of cutting parameters, the properties of the tool, and the workpiece. The effects of cutting speed, feed rate, depth of cut, shape of the cutting edge, tool wear, tool coating, and workpiece hardness on residual stresses are investigated separately.

The residual stress is measured by a variety of techniques, such as X-ray diffraction [14, 15] method, hole drilling method [16, 17], and deflection-etching method [18]. The research materials include steel [19–21], ceramics [22, 23], and composites [24].

The hardness of the workpiece has a decisive effect on the distribution of residual stress on the surface of workpiece, which is mainly manifested by affecting the shear angle in the machining process. Experimental results indicate that the compressive residual stress is easily generated on the surface of the workpiece under large shear angle condition [25]. Liu and Barash [26] observed that the length of the shear plane determines the depth of the residual stress in the workpiece, while the shape of the cutting edge, such as the edge radius and effective clearance angle, determines the residual stress distribution pattern at the machined surface. A similar conclusion regarding the dominant effect of tool nose radius on residual stress at the surface was reached by Liu et al. [27] during hard turning of bearing steel JIS SUJ2. The influence of tool edge geometry on residual stress was also confirmed by Thiele et al. [28] during hard turning of AISI 52,100 steel and by Arunachalam et al. [29] in turning of Inconel 718 alloy. It is indicated that increasing nose radius lead to tensile residual stresses. The aggravation of tool wear will increase the tensile residual stresses at the machined surface, and the compressive residual stress beneath the machined surface. Increasing cutting speed, another major influence, is observed to result in tensile residual stresses in experiments using tungsten carbide tool on AISI 316L steel [30], WC tool on gamma TiAl [31] or cubic boron nitride tool on Inconel 718 alloy [29]. Experimental results showed that the magnitude of residual stress and the depth of affected layer change with the cutting speed. With the increase of cutting speed, the absolute value of compressive residual stress on the surface decreases, and the depth of the affected layer decreases as well. In high speed milling experiment, it is reported that the feed rate has a strong influence on the shape of the residual stress profile. In the feed direction, the compressive residual stress on the machined surface increases with the increase of cutting speed. While in the direction perpendicular to the feed, the surface tensile residual stress is observed [32]. In the experiment of cutting Inconel718 alloy with CBN tool, it was found that the value of residual stress on the machined surface decreased with the increase of cutting depth [29], which was inconsistent with the previous conclusion that the cutting depth had little influence on residual stress [33]. Zhang et al. [34] conducted the turning test of low carbon steel and found that with the increase of cutting depth, the depth of the affected layer of residual stress increased significantly. The feed rate and the cutting speed had a great influence on the residual stress at the machined surface. With the increase of the feed rate and the cutting speed, the tensile residual stress on the machined surface increased as well. Hua et al. [35] conducted the hard-turning experiment of bearing steel AISI52100, and found that the feed rate had a great influence on the distribution of residual stress. With the increase of the feed rate, the maximum compressive residual stress and its depth also increase. Arunachalam et al. [29] conducted turning tests of Inconel718 alloy with CBN and ceramic tools. It is found that the use of cutting fluid in the cutting test made the surface of the workpiece prone to produce compressive residual stress. Attanasio et al. [36] studied the influence of lubrication conditions on residual stress through turning test of AISI1045, and found that lubrication effect had no significant influence on residual stress. Qin et al. [37] studied the influence of prestress on the machined residual stress, and found that the larger the prestress value was, the larger the compressive residual stress was.

1.2.2 Finite Element Research of the Residual Stress

The finite element method (FEM) based simulation techniques offer good predictions without the necessity of doing a series of experiments. However, FEM method will take excessive amounts of time when changing process parameters and repeating the analysis in order to measure the sensitivity of processes to certain parameters. Their predictions are generally close to the experimental values, especially in the measurement of residual stresses once an accurate material constitutive model is known. For this reason, FEM models continuously focus on calibrating material constitutive model (often the Johnson–Cook material model) parameters or flow stress data using experimental results and simulation outputs. With the use of better material constitutive models and elastic–viscoelastic deformations combined with the FEM, computational software has become available to obtain solutions for a rich set of field variables, providing much detailed insight for the chip formation processes. These FEM-based simulation techniques offer a detailed analysis of the physical process variables based on continuum mechanics principles on computational ground; hence they are fairly accurate in representing the physical phenomenon that take place during machining processes.

Utilizing these simulations in 2D analysis, it is possible to predict the distributions of strains, temperatures and stresses, and residual stresses after relaxation, and white layer thickness. These predictions generally use elastic, plastic, elastic–plastic, or elastic–viscoelastic models in order to simulate the machining conditions in best way, and they use 2D orthogonal cutting results in their models, as well as workpiece and tool material properties.

Since most of industrial machining operations are 3D operations, and it is necessary to extract the strain, stress, and temperature distributions in the 3D cutting process through spatial transformation. So as to predict the residual stress distribution, micro-hardness and white layer thickness of the actual machined surface by these process variables, and even predict the tool wear amount. With the rapid development of computer and numerical methods, finite element technology of 3D cutting has been developed rapidly. The chip forming process is simulated based on the damage criterion and automatic mesh redivision technology is adopted in 3D cutting simulation process. Although some problems in 3D finite element simulation have not been solved, most of the finite element research focuses on the 3D simulation of the cutting process. For a long time, the finite element method cannot accurately simulate the chip separation process from the workpiece, and it mostly relies on the manual chip separation method. It is important to note that the modified Lagrange finite element formula and mesh reconstruction technique can be used to simulate the plastic flow process of material near the tool tip, resulting in continuous or discontinuous chip formation, which avoids the simplified model of artificial chip separation criterion. In addition, some researchers used Arbitrary Lagrangian Eulerian (ALE) finite element analysis to simulate the rectangular cutting and 3D oblique cutting experiments under 2D plane strain conditions thereby advancing the state-of-the-art towards more realistic FE models for practical industrial machining operations.

Cutting force, temperature, residual stress and white layer thickness predictions have been presented in machining alloy steels such as AISI1045, AISI316, AISI52100, AISI4340 etc. These studies present fairly accurate predictions of residual stress and white layer thickness in steel machining, especially in hard turning of AISI52100 steel. Chen et al. [38] predicted the residual stress using ABAQUS software with the Johnson–Cook model and its damage model for segmental chip formation was also investigated. They also investigated the effects of chip formation on the residual stress predictions and concluded that the predictions using the damage module for segmental chips improved the residual stress predictions. Calamaz et al. [39] developed a modified material model for Ti6AlV4 alloy with considering the strain softening in serrated chip formation process. Ozel et al. [40] developed a modified material model with temperature-dependent flow softening model for Ti6Al4V alloy and investigated the effects of tool coatings on the forces, and distributions of temperatures, stresses and tool wear rate during serrated chip formation using elastic-viscoplastic 2D and visco-plastic 3D simulation models. Qin et al. [37] established a finite element model of prestressed cutting by Deform software, and deduced a prediction formula of residual stress on the machined surface as a function of prestress and cutting parameters through unitary linear regression analysis. Huang et al. [41] and Wang et al. [42, 43] simulated the cutting process of 40CrNiMoA, and investigated the key techniques in the machining process simulation, such as chip separation criterion and friction model of chip interface. The geometric-stress chip separation criterion was proposed and the residual stress distribution on the machined surface was obtained. Liu et al. [44] established a 3D finite element model by using ABAQUS software. The effects of tool geometry and cutting parameters on cutting force, cutting temperature, and machined residual stress were studied by the finite element simulation process. Zhang et al. [45, 46] simulated the hard turning of bearing steel AISI52100 by using ABAQUS software. The influence of tool edge radius on the residual stress of workpiece surface was studied, and the influence of mechanical stress and thermal stress on the residual stress were considered respectively.

1.2.3 Analytical Modeling of Residual Stress

Compared to the finite element method, analytical models have advantages by significantly reducing computational time from days to seconds, and with a more accurate and rapid applicable form for machining process optimization in industry. Indeed, this method is not only allowing the prediction of residual stress quickly and precisely, which is a crucial factor from the engineering perspective. But also, from the scientific perspective, it provides a deeper understanding of the effects of each parameter on the machined residual stress, and thus enables the optimization of the entire process as well [47].

The early effort of analytical modeling can be tracked back to Merwin and Johnson's work analyzing plastic deformation during repeated rolling contact on an elastic-perfectly plastic material under plane strain and predicting the variation of residual stresses with the distance from the contact surface [48]. Barash and Schoech [49] predicted the residual stress present in the subsurface layer of a workpiece using a simple slip-line field model. Subsequently, Liu and Barash [50, 51] studied the mechanical behavior of the machined surface with a tool with flank wear. Wu and Matsumoto [25] used an analytical model to study the effect of workpiece hardness on the pattern of residual stress induced in the machined surface. Fuh and Wu [52] developed a mathematical model for predicting the residual stress generated from end-milling of 2014-T6 alloy. The relationships between the residual stress and the cutting parameters were established through curve fitting. Although good correlations between the experimental and predicted results were obtained, the model provided little insight into the physical relationship between the cutting conditions and the residual stresses. Subsequently, Jiang and Sehitoglu [53] studied the residual stress and progressive shear strain ratcheting for the line rolling contact problems utilizing a semi-analytical approach in conjunction with a new plasticity model. As an alternative, McDowell [54] proposed an approximate algorithm for two-dimensional elastic–plastic rolling/sliding line contact which admits arbitrary forms of kinematic hardening models for non-proportional cyclic plasticity. Umbrello et al. [55] proposed a predictive model based on the artificial neural network approach that can be used both for forward and inverse predictions. The influence of material microstructure changes on residual stress distribution is also considered in the prediction model. Quteiro et al. [56] investigated the effect of cutting-edge radius on the residual stress with using heat partition method and chip flow prediction model. Ulutan et al. [57] established an analytical prediction model of the residual stresses. A finite difference-based method was utilized in the heat balance equations for determining the thermal fields of the tool, the chip, and the workpiece. The thermal field of the workpiece was used in the thermo-mechanical model of the residual stresses. Stresses resulting from thermal and mechanical loading were calculated using an analytical elastic–plastic model and a relaxation procedure. Lazoglu et al. [58] developed an enhanced analytical model by using the superposition of thermal and mechanical stresses on the workpiece, followed by a relaxation procedure. Liang and Su [59] developed an analytical model for predicting residual stresses in orthogonal cutting. The prediction model is established through the hybrid algorithm proposed by McDowell. Based on this work, Liang et al. [60] subsequently developed a physics-based model to quantitatively suggest the cutting condition and tool geometry parameters according to pre-specified surface residual stress resulting from machining. Later, Su et al. [61] extended the residual stress model to the milling process.

So far, most of the residual stress analytical modeling were focused on dry machining, few of them were concerned on the cutting fluids effect on residual stress. The influence of cutting fluid on the residual stress of workpiece surface is mainly manifested in two aspects: first, the contact stress in machining process is changed by varying the friction coefficient; second, the thermal stress in the machining process is changed by the cooling effect. Therefore, it needs to consider both the lubrication and cooling effects of MQL in residual stress modeling.

In metal cutting process, the cutting temperature determines the chip shape, the surface integrity of the workpiece, and the tool wear. The heat generated in the machining process and its heat partition coefficient have an important effect on the temperature distribution of the chip, the tool, and the workpiece. Therefore, studying the temperature distribution of the chip, the tool, and the workpiece during cutting helps to understand the tool failure mechanism and the surface integrity of the workpiece. According to the existing literature review, the experimental method or finite element method (FEM) mainly focuses on the influence of tool geometry, material properties, and cutting parameters on the cutting temperature distribution. There are few studies on the theoretical modeling of cutting temperature in the cutting process.

Moufki et al. [62] established a mathematical model for temperature prediction under orthogonal cutting conditions by considering the variation of friction coefficient with temperature. The average temperature at a certain point in the machining process is calculated according to the friction coefficient, shear angle, and some other parameters. And the temperature and stress distribution can also be obtained.

Trigger and Chao [63] established a steady-state two-dimensional analytical model to predict the average temperature during metal cutting, which can predict the average temperature at the chip-tool interface. The heat source at the shear plane and the tool-chip interface is assumed to be evenly distributed. In addition, it is assumed that the thermal shear strain energy is no longer redistributed when the contact time between the tool and the chip is very short. They calculated the thermal energy distribution of the shear plane using the Blok distribution principle [64]. Both the raw surface and the finished surface are assumed to be adiabatic. The heat partition coefficient at the tool-chip interface is calculated to predict the average temperature at tool-chip interface. Similar work was also done by Loewen and Shaw [65]. Later, Chao and Trigger proposed another temperature prediction method. They assumed the heat flux as an exponential function, which more closely approximates a true interfacial temperature distribution. In addition, they also proposed a discrete numerical iteration method, which combines the analytical method with numerical analysis method. In this way, the temperature distribution at the tool, the chip, and the interface can be predicted using Jaegaer's moving and stationary heat source solutions.

Radulescu and Kapoor [67] proposed a 3D analytical model to predict the cutting temperature in continuous and discontinuous cutting processes. This analytical modeling requires the input of cutting forces to predict the transient cutting temperatures during the cutting process.

Komanduri and Hou [68–70] developed an analytical model to predict the temperature distribution at the tool-chip interface during cutting process. The model considers both the shear heat source generated by the shear surface and the friction heat source at the tool-chip interface. The model is based on the oblique moving shear heat source proposed by Hahn [71], the frictional heat source at the tool-chip interface proposed by Chao and Trigger [66], and the moving and stationary heat source solution proposed by Jaeger. They assumed that only the back of the chip is adiabatic, and the tool-chip interface has a non-uniformly distributed moving heat source zone on one side and a rectangular heat source zone on the other side. The model is used to predict the temperature of ordinary steel processed by carbide tools.

Huang and Liang [72] considered the influence of tool wear on cutting temperature during machining on the basis of Komanduri's research. Subsequently, Li and Liang [73] extended Komanduri's analytical model for temperature prediction to near dry machining (NDM), considering the influence of cutting fluid on cutting temperature. However, the prediction model relies on the measurement of cutting force during the experiment, and the cutting temperature can be predicted according to the measured cutting force. Ji et al. [74] proposed a new method to verify the temperature prediction model based on Komanduri's temperature prediction model. The method combines the hardness of the workpiece surface with the cutting temperature history, and calculates the temperature history in the cutting process according to the measured hardness distribution of the machined surface.

The existing temperature prediction model has two disadvantages: (1) the influence of cutting fluid on cutting temperature is not considered, and the prediction model is only suitable for dry cutting condition; (2) The influence of cutting fluid on cutting temperature is considered, but the prediction of cutting temperature depends on the measurement of other physical parameters. For example, it needs to measure the cutting force and then use the measured cutting force to predict the cutting temperature.

Cutting force analysis is the premise of cutting temperature prediction, tool life prediction, cutting process optimization, and chatter analysis. In the past few decades, extensive analysis and research were conducted on the machining mechanism and cutting force in metal cutting process. According to the existing literature, cutting force modeling is mainly based on the two basic principles. One is the minimum energy principle, and the other is slip line field theory.

Merchant [75] first proposed a cutting force prediction model in orthogonal cutting process based on the principle of minimum energy. The model assumes that the workpiece material is rigid-perfectly plastic, and the thermal effect is ignored in the model. the friction coefficient at the tool-chip interface is assumed to be a constant. The cutting force varies with the shear angle, which is determined by the minimum cutting force. Similarly, Matsumura and Usui [76] proposed a cutting force prediction model for the complex tool cutting condition based on the energy method, which can be used to predict cutting forces in ball milling and rough end milling processes.

Oxley [77] established a thermo-dynamic coupling prediction model for orthogonal cutting based on slip line field theory. In this model, the temperature effect and the material properties of the workpiece are considered comprehensively. Large strain rate and strain hardening in cutting process are considered as well. A thermal analysis model is established based on experimental observation. The model prediction results agree well with the measured results under the conventional cutting speed condition. However, since the prediction model is based on the experiments, some parameters need to be calibrated by the measured results. Subsequently, Oxley's prediction model was applied to some other materials other than steel [78]. Johnson–Cook material constitutive model [79, 80], power law history material constitutive model [81], and mechanical critical stress material constitutive model [82] are commonly used in combination with Oxley's model to predict the flow stress of materials and thus predict the cutting forces in the machining.

Karpat and Ozel [80] proposed a new analytical model to predict the non-uniformly distributed heat source density at the tool-chip interface by modifying Oxley's prediction model.

Li and Liang [83] extended Oxley's prediction model to near-dry machining, and the influence of air-oil mixture on cutting force is considered. The prediction model combined Oxley's prediction theory for sharp tool and Waldorf's prediction model [84] for worn tool. It can predict cutting forces in near-dry cutting process considering tool wear. However, it needs to measure the cutting temperature as the model input.

1.3 Literature Review of Minimum Quantity Lubrication Machining

MQL machining can not only significantly reduce the cost of cutting fluid, but also minimize the harm of cutting fluid to the environment and human body. Therefore, as a viable alternative to traditional flood cooling machining, MQL machining has been more and more widely used in industrial production. According to the existing literature review, the current research on MQL machining can be divided into: (1) experimental research of MQL machining, which investigates the lubrication effects on cutting force, cutting temperature, surface roughness, and tool wear by experimental observation; (2) analytical modeling of MQL machining, which establishes the prediction model of cutting force, cutting temperature, and air quality based on the physical model. The state of art and new developments of MQL machining are discussed in the following.

1.3.1 Literature Review of MQL Experimental Investigation

The experimental research of MQL machining mainly focus on the air-oil mixture effect on the cutting force, cutting temperature, tool wear, chip morphology, quality of the machined surface. The materials of MQL machining include ordinary carbon steel, hardened steel, stainless steel, high temperature alloy etc. The processes of MQL machining mainly include turning, milling, grinding, drilling, etc.

Dhar et al. [85, 86] conducted the MQL turning experiment of AISI1040, and found that compared to dry machining, the cutting force was reduced by 5~15% in MQL machining, especially in the cutting direction. It is believed that the tribology properties at the tool-chip interface was changed by MQL, and thus the cutting temperature decreases. This is the main reason for the reduction of cutting force. In addition, the reduction in cutting temperature due to MQL mitigates the tool wear and thus extends the tool life.

Hwang and Lee [87] compared the cutting force and surface roughness of AISI1045 under MQL and traditional flood machining. Results showed that the quality of the machined surface under MQL was significantly better than that under traditional flood machining. In addition, a set of optimized parameters of cutting speed, feed rate, depth of cut, and the nozzle size was obtained by response surface analysis method.

Liu et al. [88] and Cai et al. [89] conducted the MQL milling experiment of Ti6Al4V alloy, and investigated the influence of MQL parameters such as air pressure, flow rate of cutting fluid, nozzle position, etc. on cutting force and milling temperature. The effects of cutting fluid flow rate on cutting force, surface roughness and tool wear were analyzed. It is found that the cutting force and surface roughness can be reduced within a certain range of cutting fluid flow rate. While the flow rate exceeds a certain range, the flow rate of cutting fluid on cutting force and surface is not more significant.

He's group [90–93] investigated the high speed turning and milling experiments of Titanium alloy, high temperature alloy and high strength stainless steel under different lubrication conditions. Results showed that the friction coefficient at tool and machined surface interface is significantly reduced under cryogenic MQL and MQL condition. The temperature of cryogenic MQL machining is lower than the temperature of MQL machining. In addition, the cryogenic MQL machining has no negative effect on the quality of the machined surface, and the work hardening phenomenon is not serious.

Yuan's group [94–97] studied the tool wear and chip morphology in MQL milling of high strength steel. Compared to the dry milling, it is found that the cryogenic MQL machining can effectively inhibit the formation of adhesives at the blade and prolong the tool life.

Li's group [98–101] investigated the application forms, process characteristics, and cooling and lubrication mechanism of cryogenic MQL technology. The cutting process ability of the ionic liquid in cryogenic MQL is analyzed. Cold air nano-fluid MQL machining was proposed based on the combination of excellent lubrication effect of nano-fluid and the good cooling effect of low temperature air.

Wang's group [102, 103] studied the action mechanism of cutting fluid in turning of 45 steel. The effects of MQL parameters and nozzle position on cutting force, deformation coefficient and chip breaking were analyzed, and the determination of the optimal cutting fluid amount in MQL machining was discussed.

Wang's group [104–106] analyzed the mechanism, key equipment and process applications of various MQL. The performance and adjustment of the oil was investigated under different MQL parameters and mist particle transport modes. The friction and wear properties of S136 hardened abrasives machined with AlTiN coating tool under MQL condition were simulated.

Xu's group [107–109] proposed a new green cutting technology of electrostatic spray MQL, which improved the lubrication and cooling performance of atomized cutting fluid by utilizing the characteristics of electrostatic spray charged droplets with small particle size, reduced surface tension, good adsorption and ionic air flow. The cutting characteristics and the lubrication and cooling performance of electrostatic spray MQL were compared to the traditional flood lubrication machining by experimental investigation, as well as the influence of electrostatic spray cutting fluid on the cutting environment air quality.

Zhang's group [110–112] studied the cooling and lubrication effects on the cutting force and cutting temperature in high-speed cryogenic MQL milling of 300 M steel. The prediction model of milling force and milling temperature was established as function of cutting parameters. The tool wear model and the surface roughness model were established, and the distribution law of work hardening and residual stress were summarized as well.

1.3.2 Literature Review of MQL Analytical Modeling

The research on the analytical modeling of MQL machining mainly focuses on the analytical modeling of cutting force, cutting temperature, and air quality.

Ji and Liang [113–115] used boundary lubrication theory and Jaeger's moving and stationary heat source solution to analyze the lubrication and cooling effects on cutting force, cutting temperature, and residual stress. The prediction model was developed as function of cutting parameters, tool geometry, and lubrication conditions. The sensitivity analysis of residual stress on cutting parameters, tool geometry, and lubrication condition were discussed as well.

Hadad and Sadeghi [116] established a thermal analysis model under MQL grinding by analytical method. The relationship between the flow rate of MQL and cooling heat conversion coefficient in grinding process was studied and verified by experiments.

Li [117, 118] established the relationship between the flow rate of cutting fluid, nozzle direction, nozzle size and cooling surface thermal conversion coefficient through analytical modeling. The prediction models of thermal convective coefficient under two cooling conditions were analyzed. Results showed that the cooling effect of the cutting fluid applied between the tool flank surface and the machined surface is more significant than the cooling effect applied on the back of chip.

Yan et al. [119] established a mathematical model of penetration length of MQL based on capillary theory. With this model, the influence of different cutting parameters on lubrication in vibration-assisted machining was studied. The time-varying size of microchannel between tool and workpiece was analyzed. Results showed that the filling length mainly depends on the physical properties, cutting parameters, and vibration parameters of lubricants. The predicted capillary penetration length showed good agreement with the experimental results.

Bayat et al. [120] estimated the energy consumption of each section according to the power characteristics and parameters extracted from the numerical control code, and then calculated the total energy consumption by adding the energy consumption of all parts of the machine tool. The energy consumption of milling processes was measured and compared under conventional flood machining and MQL machining. The proposed method is verified by comparing the estimated energy consumption with the experimental results.

Sofyani [121] evaluated the heat generated in the cutting process by comprehensive analysis, and established an analysis model to predict the heat generated in the shear zone and the heat distribution of the shear surface in the milling process. In addition, the heat transfer coefficients of cutting fluids in conventional flood machining, MQL machining, and spindle cooling (TSC) strategies were evaluated.

Through the analysis of the research status of MQL machining, it was found that the current research on MQL technology is mostly limited to qualitative analysis of the influence of minimum quantity lubricates on cutting performance through the experimental method, such as the influence of cutting force, cutting temperature, workpiece surface roughness, tool wear, chip morphology, etc. However, there is a lack of systematic scientific quantitative research on the machining mechanism and the cutting performance. In particular, quantitative analysis and research on MQL parameters, such as the flow rate of cutting fluid, air pressure, air-oil mixture ratio, etc., are not systematic and comprehensive.

In addition, there are few researches on residual stress under MQL machining conditions, and researches are only limited to experimental qualitative analysis, lacking of quantitative researches on its mechanism. The residual stress on the machined surface is very important to the quality of the workpiece, which will seriously affect the service life of the workpiece. Therefore, it is necessary to establish a prediction model of residual stress on the cutting surface of MQL based on the analysis of MQL machining mechanism, and quantitatively analyze the variation of residual stress on the workpiece surface with cutting parameters, tool geometric parameters, material properties and lubrication conditions. In a word, the study of residual stress in MQL machining is an inevitable trend of MQL research.

1.4 Motivation of the Research

The objective of this book is to develop a new methodology to predict the residual stress in MQL machining, to investigate how the residual stress vary with lubrication conditions, and to validate the prediction model by experimental data.

In this book, the effects of MQL on friction coefficient and heat transfer coefficient are analyzed by physics-based method. For the cooling effect, the heat loss method is adopted to predict the temperature distribution; for the lubrication effect, the boundary lubrication model is used to predict the friction coefficient under MQL condition. Both the cooling and lubrication effects are coupled in the modified Oxley's model to predict the cutting force and temperature in MQL machining. The obtained cutting force and temperature are then incorporated into a thermal–mechanical model to forecast the machined-in residual stresses in MQL machining by the hybrid algorithm which was proposed by McDowell. The proposed prediction model is verified by orthogonal turning experiments of AISI 4130 alloy steel in terms of cutting force, cutting temperature and residual stress.

In this book, the mechanical and thermal loads are incorporated in the kinematic hardening and blending function analysis to predict the residual stress in MQL machining. The proposed model can predict the residual stress profiles generated by machining directly from the process parameters, material properties, and lubrication conditions. Therefore, it provides an advantage over previous efforts because it does not require any extensive measurement of cutting force and cutting temperature.

1.5 Arrangement of Chapter

The structure of the book is as follows:

This chapter introduced the background and literature review of MQL machining. The motivation and the arrangement of chapters are also introduced.

Chapter 2 analyzed the influence of MQL on friction characteristics in machining. The penetration mechanism of cutting medium under MQL and traditional flood cooling machining is compared. The lubrication and cooling effects of MQL on cutting process were discussed respectively. The lubrication effect on the friction coefficients at the interface between chip and tool as well as the surface between tool and workpiece under MQL condition were studied based on the boundary lubrication model. The cooling effect on the heat transfer coefficient in cutting process was studied based on the forced convection cooling model.

Chapter 3 developed a thermo-mechanical coupling cutting force and cutting temperature prediction model. The thermal–mechanical coupling effect is considered from two aspects. First, the thermo-mechanical coupling cutting force and cutting temperature prediction model is established based on the material flow stress model. Second, the cutting force and cutting temperature are calculated by coupling iterative method.

Chapter 4 established the residual stress prediction model with considering the thermo-mechanical effect. The Hertz rolling contact model was used to calculate the stress distribution on the workpiece surface due to mechanical load and thermal load. Based on the modified McDowell hybrid algorithm, the residual stress distribution on the surface of the workpiece was predicted by a thermo-elastic–plastic incremental plastic model.

Chapter 5 introduced the prediction model validation. The turning experiments of AISI 4130 steel under dry, MQL and flood cooling conditions were conducted.

The predicted cutting force, cutting temperature and residual stress were compared with the measured results.

Chapter 6 discussed the sensitivity analysis of cutting performance. Based on the validated prediction model, the effects of MQL parameters such as flow rate of cutting fluid and the air-oil mixture ratio, the cutting parameters such as cutting speed, feed rate and depth of cut, the tool geometry such as rake angle and tool edge radius on the cutting force, cutting temperature, and residual stress were analyzed.

References

1. Field M, Kahles JF (1964) The surface integrity of machined and ground high-strength steels. In: Defense metals information center report, Columbus, OH, pp 54–77
2. Ding WW, Gao XL, Liu J (2022) Research status of residual stress detection methods. J Baoji Univ Arts Sci (Soc Sci Edition) 42(01):103–108 (in Chinese)
3. Wang N, Luo L, Liu Y, Peng RY, Wu J (2017) Research progress on stress measurement technology for metal components. Chin J Sci Instrument 38(10):2508–2517 (in Chinese)
4. Zhang S, Li BX, Li QH, Man J (2019) Research development of application of finite element method on cutting process. Aeronaut Manuf Technol 62(13):14–28 (in Chinese)
5. Wang SQ, Li JG, He CL, Laghari R (2019) An analytical model of residual stress in orthogonal cutting based on the radial return method. J Mater Process Technol 273:116234
6. Ji X, Li B, Liang SY (2018) Analysis of thermal and mechanical effects on residual stress in minimum quantity lubrication (MQL) machining. J Mech 34(1):41–46
7. Niosh (1998) What you need to know about occupational expose to metalworking fluids. DHHS (NIOSH) Publication, Cincinnati, OH, pp 98–116
8. Thornburg J, Leith D (2000) Mist generation during metal machining. J Tribol 122(3):544–549
9. Zhang T, Ruan JX, Cheng W (2020) Progresses in the treatment processes and techniques for cutting fluid wastewater. Chin J Environ Eng 14(9):2362–2377 (in Chinese)
10. Chen M, Peng R, Zhao LF, Chen R, Luo Y (2022) Effects of minimum quantity lubrication strategy with internal cooling tool on machining performance in turning of nickel-based superalloy GH4169. Int J Adv Manuf Technol 118(11):3673–3689
11. Akhtar W, Lazoglu I, Liang SY (2022) Prediction and control of residual stress-based distortions in the machining of aerospace parts: a review. J Manuf Process 76:106–122
12. Jawahir IS, Schoop J, Kaynak Y, Balaji AK, Ghosh R, Lu T (2020) Progress toward modeling and optimization of sustainable machining process. J Manuf Sci Eng 142:110811
13. Shao Y, Fergani O, Li B, Liang SY (2016) Residual stress modeling in minimum quantity lubrication grinding. Int J Adv Manuf Technol 83(5–8):743–751
14. Luo Q (2022) A modified X-ray diffraction method to measure residual normal and shear stresses of machined surfaces. Int J Adv Manuf Technol 119(5):3595–3606
15. Zhang R, Li A, Song X (2022) Surface quality adjustment and controlling mechanism of machined surface layer in two-step milling of titanium alloy. Int J Adv Manuf Technol 119(3):2691–2707
16. Jiang H, Wang C, Ren Z, He L (2021) Comparative analysis of residual stress and dislocation density of machined surface during turning of high strength steel. Procedia CIRP 101(15):38–41
17. Schajer GS (2022) Universal calibration constants for strain gauge hole-drilling residual stress measurements. Exp Mech 62(2):351–358
18. Ebrahimi A, Bayat M, Norouzi E (2021) Measurement of residual stress using the ultrasonic method in aluminum welding: FE analysis and experimental study. Russ J Nondestr Test 57(8):669–678

19. Chavan A, Sargade V (2020) Surface integrity of AISI 52100 steel during hard turning in different near-dry environments. Adv Mater Sci Eng 1–13
20. Masmiati N, Sarhan AA, Hassan M, Hamdi M (2016) Optimization of cutting conditions for minimum residual stress, cutting force and surface roughness in end milling of S50C medium carbon steel. Measurement 86:253–265
21. Kulecki P, Lichańska E (2017) The effect of powder ball milling on the microstructure and mechanical properties of sintered Fe-Cr-Mo-Mn-(Cu) Steel. Powder Metallurgy Progress 17(2):82–92
22. Guo WC, Li B, Shen S, Wu C, Hu J (2018) An empirical model for prediction of residual stress based on grinding forces. Mater Sci Forum 939:46–53
23. Ling H, Yang C, Feng S, Lu H (2020) Predictive model of grinding residual stress for linear guideway considering straightening history. Int J Mech Sci 176:105536
24. Clau B, Nestler A, Schubert A, Dietrich D, Lampke T (2019) Investigation of surface properties in turn milling of particle-reinforced aluminium matrix composites using MCD-tipped tools. Int J Adv Manuf Technol 105(143):937–950
25. Wu D, Matsumoto Y (1990) The effect of hardness on residual stresses in orthogonal machining of AISI 4340 steel. J Eng Indus Trans ASME 112(3):245–252
26. Liu C, Barash M (1982) Variables governing patterns of mechanical residual stress in a machined surface. J Eng Indus 104(3):257–264
27. Liu M, Takagi J, Tsukuda A (2004) Effect of tool nose radius and tool wear on residual stress distribution in hard turning of bearing steel. J Mater Process Technol 150(3):234–241
28. Thiele J, Melkote SN, Peascoe RA, Watkins TR (2000) Effect of cutting-edge geometry and workpiece hardness on surface residual stresses in finish hard turning of AISI 52100 steel. J Manuf Sci Eng 122(4):642–649
29. Arunachalam R, Mannan M, Spowage A (2004) Residual stress and surface roughness when facing age hardened Inconel 718 with CBN and ceramic cutting tools. Int J Mach Tools Manuf 44(9):879–887
30. M'saoubi R, Outeiro JC, Changeux B, Lebrun JL, Dias AM (1999) Residual stress analysis in orthogonal machining of standard and resulfurized AISI 316L steels. J Mater Process Technol 96(1):225–233
31. Mantle AL, Aspinwall DK (2001) Surface integrity of a high speed milled gamma titanium aluminide. J Mater Process Technol 118(1):143–150
32. Mu YJ, Guo GQ, Hu M, Ming WW, Chen M (2018) Analysis of surface quality for tank panel material under high-speed milling. Tool Eng 52(05):60–62 (in Chinese)
33. Capello E (2005) Residual stresses in turning: part I: influence of process parameters. J Mater Process Technol 160(2):221–228
34. Zhang YL, Huang HR, Li X (2006) Research on distribution of turning residual stress. J Beijing Univ Technol 32(7):582–584 (in Chinese)
35. Hua J, Shivpuri R, Cheng XM, Bedekar V, Matsumoto Y, Hashimoto F, Watkins TR (2005) Effect of feed rate, workpiece hardness and cutting edge on subsurface residual stress in the hard turning of bearing steel using chamfer + hone cutting edge geometry. Mater Sci Eng, A 394(1):238–248
36. Attanasio A, Ceretti E, Gelfi M, Giardini C (2009) Experimental evaluation of lubricant influence on residual stress in turning operations. Int J Mach Mach Mater 6(1–2):106–119
37. Tan MY, Xu LY, Pan XL, Li YX, Liang CJ (2021) Experimental investigation on turning of stainless steel by nanofluid minimal quantity lubrication. Mach Tool 49(08):42–45 (in Chinese)
38. Chen L, El-Wardany T, Harris W (2004) Modelling the effects of flank wear land and chip formation on residual stresses. CIRP Ann Manuf Technol 53(1):95–98
39. Calamaz M, Coupard D, Girot F (2008) A new material model for 2D numerical simulation of serrated chip formation when machining titanium alloy Ti-6Al-4V. Int J Mach Tools Manuf 48(3–4):275–288
40. Özel T, Sima M, Srivastava AK, Kaftanoglu B (2010) Investigations on the effects of multi-layered coated inserts in machining Ti-6Al-4V alloy with experiments and finite element simulations. CIRP Ann Manuf Technol 59(1):77–82

41. Huang ZG, Ke YL, Wang LT (2004) Coupled thermo-mechanical model for metal orthogonal cutting process and finite element simulation. Acta Aeronautica ET Astronautica Sinica 25(3):317–320 (in Chinese)
42. Wang LT (2003) Study on residual stresses and distortion theory of aeronautica frame structure in the milling. Zhejiang University, Hangzhou (in Chinese)
43. Wang LT, Ke YL, Huang ZG, Xu D, Wu Q (2003) Study on residual stress produced in milling of aeronautic structure. Acta Aeronautica ET Astronautica Sinica 24(3):286–288 (in Chinese)
44. Liu H, Xie W, Sun Y, Zhang JY (2018) Investigations on micro-cutting mechanism and surface quality of carbon fiber-reinforced plastic composites. Int J Adv Manuf Technol 94:3655–3664
45. Zhang XP, Wu SF, Liu CR (2012) In Hook shaped residual stress: The effect of tool ploughing, and the analysis of the mechanical and thermal effects. In: Proceedings of the 2012 ASME international manufacturing science and engineering conference, MSEC 2012 June 4–8, 2012, Notre Dame, IN, USA, pp 277–286
46. Zhang XP, Wu SF, Wang HP, Liu CR (2011) Predicting the effects of cutting parameters and tool geometry on hard turning process using finite element method. J Manuf Sci Eng 133(4):255–262
47. Arrazola PJ, Özel T, Umbrello D, Davies M, Jawahir IS (2013) Recent advances in modelling of metal machining processes. CIRP Ann Manuf Technol 62:695–718
48. Merwin J, Johnson KL (1963) An analysis of plastic deformation in rolling contact. Proc Inst Mech Eng 177(1):676–690
49. Barash M, Schoech W (1970) In semi-analytical model of the residual stress zone in orthogonal machining. In: Proceedings of 11th international MTDR conference. Pergamon Press, London
50. Liu CR, Barash MM (1976) The mechanical state of the sublayer of a surface generated by chip-removal process—Part 2: cutting with a tool with flank wear. J Eng Indus 98:1202–1208
51. Liu CR, Barash MM (1976) The mechanical state of the sublayer of a surface generated by chip-removal process—part 1: cutting with a sharp tool. J Eng Indus 98:1192–1197
52. Fuh KH, Wu CF (1995) A residual-stress model for the milling of aluminum alloy (2014–T6). J Mater Process Technol 51(1):87–105
53. Jiang Y, Sehitoglu H (1996) Rolling contact stress analysis with the application of a new plasticity model. Wear 191(1–2):35–44
54. McDowell D (1997) An approximate algorithm for elastic-plastic two-dimensional rolling/sliding contact. Wear 211(2):237–246
55. Umbrello D, Ambrogio G, Filice L, Shivpuri R (2007) An ANN approach for predicting subsurface residual stresses and the desired cutting conditions during hard turning. J Mater Process Technol 189(1–3):143–152
56. Outeiro JC, Ee KC, Dillon OW, Wanigarathne PC, Jawahir IS (2006) Some observations on comparing the modelled and measured residual stresses on the machined surface Induced by orthogonal cutting of AISI316L steel. In: Proceedings of the 9th CIRP international workshop on modelling of machining operations, Bled, Slovenia
57. Ulutan D, Erdem Alaca B, Lazoglu I (2007) Analytical modelling of residual stresses in machining. J Mater Process Technol 183(1):77–87
58. Lazoglu I, Ulutan D, Alaca BE, Engin S, Kaftanoglu B (2008) An enhanced analytical model for residual stress prediction in machining. CIRP Ann Manuf Technol 57(1):81–84
59. Liang SY, Su JC (2007) Residual stress modeling in orthogonal machining. CIRP Ann Manuf Technol 56(1):65–68
60. Liang SY, Hanna CR, Chao RM (2008) Achieving machining residual stresses through model-driven planning of process parameters. Trans North Am Manuf Res Inst SME 36:445–452
61. Su JC, Young KA, Ma K, Srivatsa S, Morehouse JB (2013) Modeling of residual stresses in milling. Int J Adv Manuf Technol 65:717–733
62. Moufki A, Molinari A, Dudzinski D (1998) Modelling of orthogonal cutting with a temperature dependent friction law. J Mech Phys Solids 46(10):2103–2138
63. Trigger KJ, Chao BT (1951) An analytical evaluation of metal-cutting temperatures. Trans ASME 73:57–68

64. Blok H (1937) Theoretical study of temperature rise at surfaces of actual contact under oiliness lubricating conditions. Proc General Discussion Lubrication 2:222–235
65. Loewen E, Shaw M (1954) On the analysis of cutting tool temperatures. Trans ASME 76(2):217
66. Chao BT, Trigger KJ (1958) Temperature distribution at tool-chip and tool-work interface in metal cutting. Trans ASME 80(1):311–320
67. Radulescu R, Kapoor S (1994) An analytical model for prediction of tool temperature fields during continuous and interrupted cutting. J Eng Indus 116(2):135–143
68. Komanduri R, Hou ZB (2000) Thermal modeling of the metal cutting process—part I: temperature rise distribution due to shear plane heat source. Int J Mech Sci 42(9):1715–1752
69. Komanduri R, Hou ZB (2001) Thermal modeling of the metal cutting process—part II: temperature rise distribution due to frictional heat source at the tool-chip interface. Int J Mech Sci 43(1):57–88
70. Komanduri R, Hou ZB (2001) Thermal modeling of the metal cutting process—part III: temperature rise distribution due to the combined effects of shear plane heat source and the tool-chip interface frictional heat source. Int J Mech Sci 43(1):89–107
71. Hahn RS (1951) On the temperature developed at the shear plane in the metal cutting process. In: Proceedings of first US national congress of applied mechanics, pp 661–666
72. Huang Y, Liang SY (2003) Modelling of the cutting temperature distribution under the tool flank wear effect. Proc Inst Mech Eng C J Mech Eng Sci 217(11):1195–1208
73. Li KM, Liang SY (2006) Modeling of cutting temperature in near dry machining. J Manuf Sci Eng 128:416–424
74. Ji X, Kang Z, Zhang XP, Liang SY (2011) A new methodology to validate the cutting temperature theoretical model in super-finish hard machining. Adv Sci Lett 4(4–5):1561–1565
75. Merchant ME (1945) Mechanics of the metal cutting process. II. Plasticity conditions in orthogonal cutting. J Appl Phys 16(6):318–324
76. Matsumura T, Usui E (2010) Predictive cutting force model in complex-shaped end milling based on minimum cutting energy. Int J Mach Tools Manuf 50(5):458–466
77. Oxley PLB (1989) The mechanics of machining: an analytical approach to assessing machinability. First edition edits on, Ellis Horwood Ltd, New York
78. Adibi-Sedeh AH, Madhavan V, Bahr B (2003) Extension of Oxley's analysis of machining to use different material models. J Manuf Sci Eng Trans ASME 125(4):656–666
79. Lalwani D, Mehta N, Jain P (2009) Extension of Oxley's predictive machining theory for Johnson and Cook flow stress model. J Mater Process Technol 209(12–13):5305–5312
80. Karpat Y, Ozel T (2006) Predictive analytical and thermal modeling of orthogonal cutting process—Part I: Predictions of tool forces, stresses, and temperature distributions. J Manuf Sci Eng Trans ASME 128(2):435–444
81. Childs T (1997) Material property requirements for modeling metal machining. Journal de Physique (France) 7:1–6
82. Follansbee P, Kocks U (1988) A constitutive description of the deformation of copper based on the use of the mechanical threshold stress as an internal state variable. Acta Metall 36(1):81–93
83. Li KM, Liang SY (2007) Modeling of cutting forces in near dry machining under tool wear effect. Int J Mach Tools Manuf 47(7–8):1292–1301
84. Waldorf DJ, DeVor RE, Kapoor S (1998) A slip-line field for ploughing during orthogonal cutting. J Manuf Sci Eng 120(4):693–698
85. Dhar N, Islam M, Islam S, Mithu M (2006) The influence of minimum quantity of lubrication (MQL) on cutting temperature, chip and dimensional accuracy in turning AISI-1040 steel. J Mater Process Technol 171(1):93–99
86. Dhar N, Islam M, Islam S, Mithu M (2007) An experimental investigation on effect of minimum quantity lubrication in machining AISI 1040 steel. Int J Mach Tools Manuf 47(5):748–753
87. Hwang YK, Lee CM (2010) Surface roughness and cutting force prediction in MQL and wet turning process of AISI 1045 using design of experiments. J Mech Sci Technol 24(8):1669–1677

88. Liu ZQ, Cai XJ, Chen M, An QL (2011) Investigation of cutting force and temperature of end-milling Ti-6Al-4V with different minimum quantity lubrication (MQL) parameters. Proc Inst Mech Eng Part B J Eng Manuf 225(B8):1273–1279
89. Cai XJ, Liu ZQ, Chen M, An QL (2012) An experimental investigation on effects of minimum quantity lubrication oil supply rate in high-speed end milling of Ti-6Al-4V. Proc Inst Mech Eng Part B J Eng Manuf 226(A11):1784–1792
90. Bian R, Ding W, Liu S, He N (2019) Research on high performance milling of engineering ceramics from the perspective of cutting variables setting. Materials 12(1):122
91. Bian R, Ferraris E, Yang YF, Qian J (2018) Experimental investigation on ductile mode micro-milling of ZrO_2 ceramics with diamond-coated end mills. Micromachines 9(3):127
92. Qi BY (2011) Research on cooling/lubrication technology of green cutting of titanium alloy using surface micro-textured cutting tool. Nanjing University of Aeronautics and Astronautics, Nanjing (in Chinese)
93. Qi BY, He N, Li L, Zhao W, Bian R (2010) Cryogenic minimum quantity lubrication technology and its action mechanism. Mech Sci Technol Aerosp Eng 29(6):826–831 (in Chinese)
94. Yuan SM, Zhu GY, Wang L (2017) Recent progress on lubricant characteristics of minimum quantity lubrication (MQL) technology in green cutting. Chin J Mech Eng 53(17):131–140 (in Chinese)
95. Yuan SM, Song H, Lu YL (2017) Research for cutting force and torque in rotary ultrasonic drilling of carbon fiber reinforced polymers. Aeronaut Manuf Technol 16:96–102 (in Chinese)
96. Yuan SM, Zhu GY, Liu S, Yan LT (2016) Orthogonal test study on the nozzle orientation of minimal quantity lubrication with cooling air technology. Aeronaut Manuf Technol 10:64–69 (in Chinese)
97. Yan LT, Yuan SM, Liu Q (2010) Tool wear and chip formation in green machining of high strength steel. Chin J Mech Eng 46(09):187–192 (in Chinese)
98. Liu MZ, Li CH, Cao HJ, Zhang S, Chen Y, Liu B, Zhang NQ, Zhou ZM (2022) Research progresses and applications of CMQL machining technology. Chin J Mech Eng 33(5):529–550 (in Chinese)
99. Wang DX, Zhao QL, Zhang Y, Gao T, Jiang JL, Liu GL, Li CH (2022) Investigation on tribological mechanism of ionic liquid on grinding interfaces under MQL. Chin J Mech Eng 33(5):560–568 (in Chinese)
100. Jia DZ, Li CH, Zhang YB, Yang M, Wang YG, Guo SM, Cao HJ (2017) Specific energy and surface roughness of minimum quantity lubrication grinding Ni-based alloy with mixed vegetable oil-based nanofluids. Precis Eng 50:248–262
101. Jia DZ, Li CH, Zhang YB, Yang M, Zhang XP, Li RZ, Ji HJ (2019) Experimental evaluation of surface topographies of NMQL grinding ZrO_2 ceramics combining multiangle ultrasonic vibration. Int J Adv Manuf Technol 100:457–473
102. Pei HJ, Li F, Chen YY, Wang GC (2022) Atomizing spray characteristics study of minimum quantity lubrication system. Aeronaut Manuf Technol 65(07):70–76 (in Chinese)
103. Pei HJ, Shen CG, Huang J, Wang GC (2017) Flow field characteristics of the wedge zone between a major flank and a transient surface. Int J Adv Manuf Technol 92(9–12):4253–4261
104. Yang JZ, Wang CY, Yuan YH, Yuan SM, Wang XB, Liang CY, Li WQ (2022) State-of-the-art on MQL synergistic technologies and their applications. China Mech Eng 33(5):506–528 (in Chinese)
105. Yuan YH, Wang CY, Yang JZ, Liang CY, Lin HT, Li WQ (2021) Friction behavior between AlTiN-based coated tools and S136 hardened steel under MQL conditions. Tool Eng 55(9):24–29 (in Chinese)
106. Liang CL, Yuan YH, Wang CY, Li WQ, Yang JZ (2022) Study on oil mist control and oil mist particle characteristics of MQL systems. China Mech Eng 33(5):607–614 (in Chinese)
107. Huang S, Yao W, Hu J, Xu XF (2015) Tribological performance and lubrication mechanism of contact-charged electrostatic spray lubrication technique. Tribol Lett 59(2):28
108. Huang S, Tao L, Wang M, Xu XF (2018) Effects of machining and oil mist parameters on electrostatic minimum quantity lubrication–EMQL turning process. Int J Precision Eng Manuf-Green Technol 5(2):317–326

109. Guan JJ, Xu XF, Li G, Peng W (2014) Preparation and tribological properties of inclusion complex of β-cyclodextrin/dialkyl pentasulfide as additive in PEG-600 aqueous solution. Appl Surf Sci 289(15):400–406
110. Zhang HP, Wang ZJ, Liu GL (2020) Tool wear of high-speed machining 300 M steel under CMQL condition. J Harbin Univ Sci Technol 25(03):75–82 (in Chinese)
111. Zhang HP, Zhang ZS, Zheng ZY, Liu EL (2020) Tool wear in high-speed turning ultra-high strength steel under dry and CMQL conditions. Integr Ferroelectr 206(1):122–131
112. Zhang HP, Ren Y, Xue FG, Liu RH (2019) Cryogenic minimum quantity lubrication processing technology. J Harbin Univ Sci Technol 24(02):38–44 (in Chinese)
113. Ji X, Zhang XP, Liang SY (2016) Predicting the effects of cutting fluid on machining force, temperature and residual stress using analytical method. Int J Comput Appl Technol 53:135–141
114. Ji X, Liang SY (2017) Model-based sensitivity analysis of machining-induced residual stress under minimum quantity lubrication. Proc Inst Mech Eng Part B J Eng Manuf 231:1528–1541
115. Ji X, Li BZ, Liang SY (2018) Analysis of thermal and mechanical effects on residual stress in minimum quantity lubrication (MQL) machining. J Mech 34:41–46
116. Hadad M, Sadeghi B (2012) Thermal analysis of minimum quantity lubrication—MQL grinding process. Int J Mach Tools Manuf 63:1–15
117. Li XP (1996) Study of the jet-flow rate of cooling in machining Part 1. Theoretical analysis. J Mater Process Technol 62(1):149–156
118. Li XP (1996) Study of the jet-flow rate of cooling in machining Part 2. Simulation study. J Mater Process Technol 62(1):157–165
119. Yan LT, Zhang QJ, Yu JZ (2018) Analytical models for oil penetration and experimental study on vibration assisted machining with minimum quantity lubrication. Int J Mech Sci 148:374–382
120. Bayat M, Abootorabi MM (2021) Comparison of minimum quantity lubrication and wet milling based on energy consumption modeling. Proc Inst Mech Eng Part E: J Process Mech Eng 235(5):1665–1675
121. Sofyani A, Sharaf, Marinescu et al (2017) Analytical modeling of the thermal aspects of metalworking fluids in the milling process. Int J Adv Manuf Technol 92:3953–3966

Chapter 2
The Effects of MQL on Tribological Attributes in Machining

In order to establish the prediction model of cutting force and cutting temperature in MQL machining, the friction characteristics of MQL on the cutting model must be analyzed first. In the process of MQL machining, only a small amount of cutting fluid is sprayed in the cutting area by the compressed air. On one hand, a thin boundary lubrication film is formed at the interface of tool and chip and the tool and machined surface by physical adsorption or chemical reaction forming, which effectively reduce the friction coefficient, lower cutting force and reduce frictional heat. It will produce the indirect cooling effect. On the other hand, the temperature of the cutting area is significantly reduced by the convective heat transfer between the air-oil mixture and the machined surface. It will produce a direct cooling effect on the metal cutting process, which is shown in Fig. 2.1. The influence of MQL on the friction characteristics in cutting process is analyzed in details from the lubrication and cooling effects.

2.1 Penetration Mechanism of Cutting Medium

In friction contact process, lubricants generally enter the contact area through the movement of the friction surface from a larger space. According to the fluid lubrication theory, an oil wedge with static pressure can be formed when the lubricant moves from large space to small space, thus separating the two friction surfaces. Similarly, in the process of metal cutting, the cutting fluid enters the tool and chip contact surface and the tool and workpiece contact surface through the friction movement between the tool and the workpiece, thus forming the oil wedge space. However, in the actual cutting process, the direction of chip and workpiece movement relative to the tool is opposite to the direction of cutting fluid entering the cutting zone, so it hinders the formation of oil wedge. It is necessary to investigate the penetration mechanism of cutting medium in MQL machining.

© The Author(s), under exclusive license to Springer Nature Singapore Pte Ltd. 2023

X. Ji, *Minimum Quantity Lubrication Machining*,
https://doi.org/10.1007/978-981-19-7087-0_2

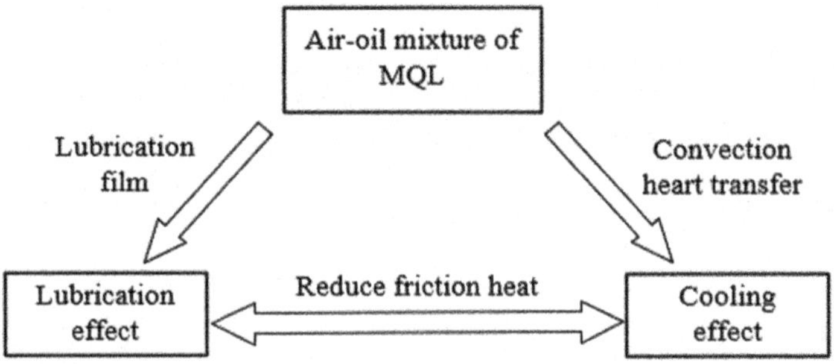

Fig. 2.1 Influence of MQL on friction characteristics

At present, there are three hypotheses on the penetration mechanism of cutting medium in metal cutting process: (1) The cutting medium enters the cutting zone by means of element diffusion through the metal lattice; (2) By adding surface active factors to the cutting medium, the cutting medium can be promoted into the cutting zone; (3) The cutting medium penetrates into the cutting friction zone through capillary action [1, 2]. The penetration mechanisms of cutting medium in traditional flood cooling and MQL machining are analyzed in the following.

2.1.1 Penetration Mechanism of Convectional Flood Cooling Machining

It is difficult for the cutting medium to enter the cutting area due to the high temperature, high speed and high pressure in cutting process. For the traditional flood cooling machining, a lot of researches were focused on the penetration mechanism of cutting medium, and most of the researchers generally agreed that the cutting medium penetrates into the friction region through the capillary action of the cutting region. Williams and Tabor proposed the hypothesis that the friction zone between cutter and chip forms a small cuboid capillary through experimental research [3]. Godelvskiy considered that the capillary of the friction zone in metal cutting was a tiny cylinder [4]. Liu conducted the plane cutting of organic glass experiments, and observed the changes of water vapor in the cutting process by using CCD computer system, and finally confirmed Godelvskiy's hypothesis. In addition, a dynamic model of cutting fluid penetrating capillary in cutting zone was established [5]. It is assumed that a large number of capillary tubes were distributed at the interface between the tool and the chip, and the capillary tubes were arranged along the direction of chip outflow. A single capillary was a cylinder with one end open and one end closed, and it is vacuum in the body. The length of the capillary is about 1/3–2/3 of the contact length of the tool and the chip, and the radius is about several to more than ten microns. It was assumed that the individual capillaries have a certain existing time t_{cell}, which is determined by the flow velocity of the chip V_c. So, the length of the capillary L_{cell} can be calculated according to the existing time t_{cell} and the chip velocity V_c.

Based on Godelvskiy's study, Liu found that in the traditional flood cooling machining, the cutting fluid permeates into the capillary by three steps: (1) liquid phase infiltration stage; (2) droplet evaporation stage; (3) gas filling stage. The state of the cutting fluid and the filling time of the capillary are different at each stage, as shown in Fig. 2.2. In the stage of liquid phase infiltration, the cutting fluid was sucked into the capillary by siphoning action. It is assumed that the length of liquid phase infiltration is L_f, the infiltration time of liquid phase is t_l. Due to the conduction of cutting heat, cutting fluid penetrates into the capillary after a distance. The internal energy of the infiltrated cutting fluid increases rapidly due to the absorption of cutting heat. The pressure of the infiltrated cutting fluid increases as well, the cutting fluid changed into water vapor due to volume expansion, as shown in Fig. 2.2b. At this time, due to the increase of pressure in the capillary, the cutting fluid at the entrance of the capillary cannot enter into the capillary, the cutting fluid inside the capillary were completely transformed into the gas phase and continued to fill the capillary, as shown in Fig. 2.2c. The filling length of the gas is defined as L_g, the filling time of the gas is defined as t_g. The total time of the cutting fluid penetrating into the capillary t_{total} is the sum of the infiltration time of liquid phase t_l and the filling time of the gas t_g. When the total penetration time of cutting fluid t_{total} is less than the existing time of the capillary t_{cell}, the cutting fluid penetrating into the capillary cannot form a lubricating film. Therefore, the total penetration time of cutting fluid directly affects the penetration and lubrication effect of cutting fluid. From the above analysis, it is known that the state of cutting fluid directly affects the lubrication effect.

2.1.2 Penetration Mechanism of MQL Machining

Different from the traditional flood cooling machining, a very small amount of cutting fluid is sprayed into the cutting area under the action of compressed air in MQL machining. In MQL machining, the air-oil mixture penetrates into the interface between the tool and chip as well as the tool and workpiece contact surface through capillary in the form of gas–liquid two phase. Compared to flood cooling machining, air-oil mixture in MQL machining does not need to go through the liquid phase infiltration and droplet evaporation stages, and directly enter into the capillary through gas phase filling, as shown in Fig. 2.3.

According to the analysis of the penetration mechanism of cutting medium in Sect. 2.1.2, the penetration time of cutting medium must be shorter than the existing time of capillary. Therefore, an important parameter to evaluate the lubrication effect of air-oil mixture in MQL machining is the filling speed of air-oil mixture in capillary. The velocity of air-oil mixture filling capillary can be determined from the momentum equation and energy equation of one-dimensional unsteady gas dynamics. For MQL machining, the total time of air-oil filling capillary t_{total} equals to the filling time of the gas t_g. The existing time of capillary is assumed to be t_{cell}. The momentum

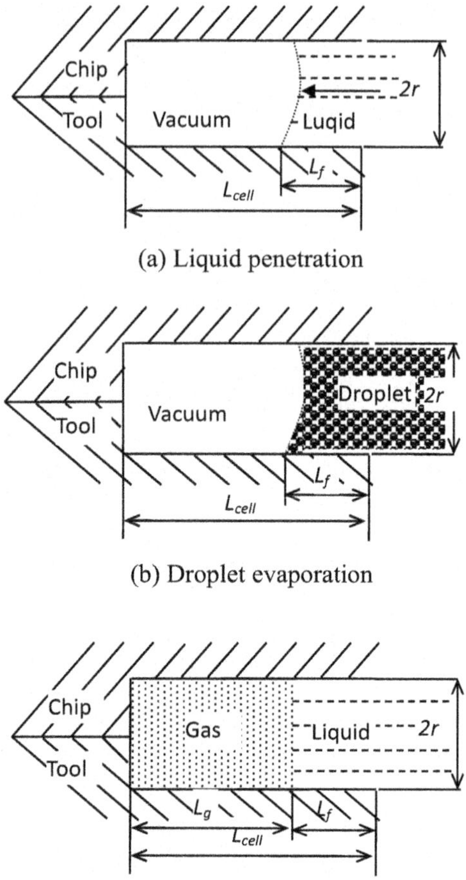

(a) Liquid penetration

(b) Droplet evaporation

(c) Gas filling

Fig. 2.2 The penetration process of cutting fluid in flood cooling machining

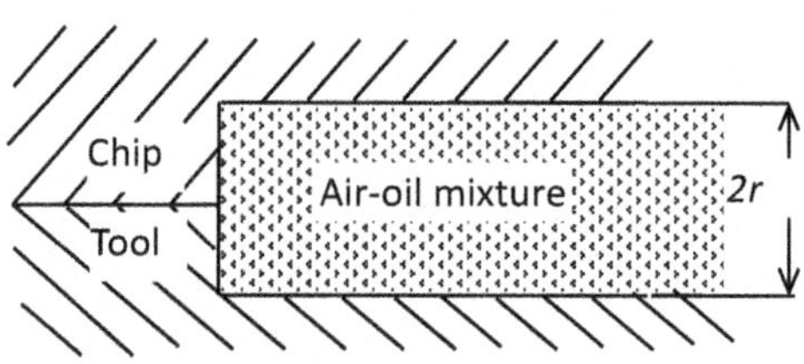

Fig. 2.3 Permeation model of MQL machining

equation of one-dimensional steady flow under the influence of friction and heating can be expressed as:

$$\rho_g \frac{\partial v_g}{\partial t} + \rho_g v_g \frac{\partial v_g}{\partial x} + \frac{\partial p_g}{\partial x} = -\rho_g v_g^2 \frac{fb}{2r} \tag{2.1}$$

where ρ_g is the density of air-oil mixture, $\partial v_g/\partial t$ is the velocity change rate of air-oil mixture filled capillary. $\partial v_g/\partial x$ is the velocity gradient of air-oil mixture filled capillary. f is the friction factor, and b is the section shape factor, for cylindrical section, b equals to two.

Considering that the air-oil mixture is affected by heat in the process of entering the capillary, the energy equation of the air-oil mixture filling the capillary can be expressed by:

$$\frac{\partial}{\partial t}\left[\rho_g\left(e+\frac{v_g^2}{2}\right)\right]Adx+\frac{\partial}{\partial x}\left[\rho_g v_g A\left(e+\frac{v_g^2}{2}\right)\right]dx+\frac{\partial}{\partial x}(p_g v_g A)dx=\delta\dot{Q} \tag{2.2}$$

where e denotes the internal energy of the air-oil mixture per unit mass, which is a function of the pressure of the compressed air and the density of the air-oil mixture. A is the cross-sectional area of the tiny unit filling capillary tube for the air-oil mixture.

It is assumed that the density of the air-oil mixture in the capillary remains constant, Eq. (2.2) can be written as:

$$v_g\frac{\partial v_g}{\partial t}\rho_g Adx+\frac{\partial v_g}{\partial x}\left(e+\frac{3}{2}v_g^2+\frac{p_g}{\rho_g}\right)\rho_g Adx+v_g\frac{\partial p_g}{\partial x}Adx=\delta\dot{Q} \tag{2.3}$$

Combing Eqs. (2.2) and (2.3), the flow velocity of air-oil mixture in capillary can be calculated by:

$$v_g=\sqrt{\frac{\frac{p_g}{\rho_g L_{cell}}+\frac{e}{L_{cell}}-\alpha 2\pi r L_{cell}\Delta T}{\frac{fb}{2r}-\frac{1}{2L_{cell}}}} \tag{2.4}$$

where α is the heat transfer coefficient between air-oil mixture droplet and air-oil mixture vapor, which is determined by the thermal physical parameters of the air-oil mixture and the size of the oil mist liquid.

Taking the water vapor as an example [5], by comparing water cooling and vapor cooling, it was found that the time of filling capillary with water vapor jet was one order of magnitude smaller than that of water, which was significantly shorter than the existing time of capillary. Therefore, under the same cutting parameters, the penetration effect of cutting medium under MQL is obviously better than that of traditional flood cooling machining, that is why the lubrication effect of MQL can reach or even exceed that of traditional flood cooling machining.

2.2 Lubrication Effect of MQL

2.2.1 Lubrication Mechanism of MQL

According to the analysis in Sect. 2.1.2, the air-oil mixture is sprayed into the cutting area through the nozzle in a gas–liquid two-phase manner in MQL machining. On one hand, under the action of siphoning, the air-oil mixture enters into the capillary of friction interface, thus forming a lubricating film to separate the two metal contact surfaces. However, due to the effects of friction and cutting heat, the pressure of air-oil mixture increases after entering the capillary, when it increases to a certain value, the capillary would crack, thus resulting in the direct contact between the two metal surfaces [6]. On the other hand, due to the minimum quantity of cutting fluid in MQL machining, which is about one in ten thousand of that in the traditional flood cooling machining, a complete lubrication film cannot be formed at the contact surface, so the lubrication film mostly exists in the form of boundary lubrication in MQL machining [7].

According to the existing research on MQL machining, it is found that the lubrication effect in MQL machining can reach or even exceed the traditional flood cooling machining. The main reasons are as follows: (1) under the traditional flood cooling machining, although a large amount of cutting fluid is applied, due to its low flow rate and small pressure, actually the amount of cutting fluid entering into the cutting area is limited, most of the cutting fluid only plays a cooling role on the surroundings, and the utilization rate of cutting fluid is not high. In addition, due to plastic deformation and friction effect in the cutting area, the cutting temperature increases suddenly, and the gas at the tool tip is heated and expanded, thus preventing the cutting fluid from entering into the tool-chip interface and the tool-workpiece interface. It significantly weakens the lubrication effect of the cutting fluid. In MQL machining, a very small amount of air-oil mixture is formed to be a small mist under high speed and high pressure, and then sprayed to the cutting area. The small mist can easily penetrate into the tool-chip interface and the tool-workpiece interface due to its small size and high speed, as well as the adjustable nozzle position. Therefore, the utilization rate of cutting fluid in MQL is high. (2) Due to the minimum amount of cutting fluid in MQL machining, and many micro rough points existing at contact surfaces, a complete lubrication film cannot be formed at the tool-chip interface and tool-workpiece interface. The boundary lubrication state exists under most cases. According to the analysis of boundary lubrication conditions, the lubrication effect does not mean that the more cutting fluid the better lubrication effect, while there is a certain range of magnitude for the cutting fluid. The range of this value is related to the properties of cutting fluid, material properties of friction pair, surface roughness, etc. The maximum amount of the cutting fluid is to fill the peak and valley of the contact area [8]. When the chip flows out along the front tool surface, there are some small gaps at the chip-tool interface, and the amount of cutting fluid required to fill the gap is not much, which is far lower than the amount of cutting fluid in the traditional flood cooling machining. (3) In cutting process, the intense friction happened at the tool-chip interface as well as the tool-workpiece interface. High-energy electrons generated at the contact surface because of extrusion and friction. The oxygen in the air-oil mixture is separated by the heat, which can be reacted with the metal material to produce an oxide film on the metal surface. It also improves the lubrication effect of MQL.

2.2.2 *Friction Coefficient in MQL*

Due to the minimum amount of cutting fluid, the complete lubrication film cannot be formed in MQL machining, and most of them exist in the form of boundary lubrication state. Therefore, the boundary lubrication model is more suitable to predict the friction coefficient in MQL machining.

Take the friction at tool-chip interface as an example. In the cutting process, when the two contact surfaces bear loads, part of the rough points will break the boundary lubrication film due to excessive contact pressure, which results in the direct contact of the surfaces, as shown the Area A in Fig. 2.4. Area B indicates the bearing area of the boundary lubrication. Area C is the lubricating area formed by rough peak points, where the boundary films of the two metal surfaces do not contact each other, so no load exists at Area C. The thickness of the boundary lubrication film is denoted as t_b. The load in the normal direction is carried by the contact area of metal A_{ms} and the absorbed lubrication film A_{bs}. The contact stress at the contact area of metal is p_m. The contact stress at the absorbed lubrication film varies with the contact area of absorbed lubrication film, which approximately equals to the yield stress at the junction of lubrication film and metal contact area. The area outside the absorbed lubrication film, such as Area C, is considered to be free of load. Therefore, the average contact stress p_b in the region of absorbed lubrication film is approximated to be the constant $C_2 p_m$.

In Fig. 2.4, the total contact area of metal is denoted as A_{ms}, the total area of absorbed lubrication film is denoted as A_{bs}. The normal load at this area can be expressed by:

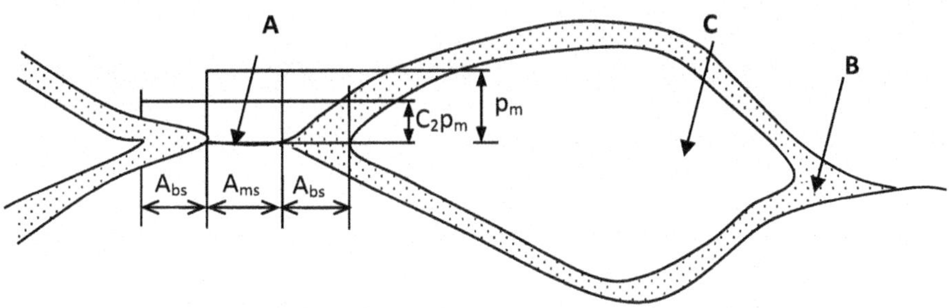

Fig. 2.4 The boundary lubrication model

$$N = p_m A_{ms} + p_b A_{bs} \tag{2.5}$$

The shear strength of metal contact area and absorbed lubrication film area are respectively represented as s_m and s_b. Therefore, the friction force can be calculated by:

$$F = s_m A_{ms} + s_b A_{bs} \tag{2.6}$$

The friction coefficient can be expressed by:

$$\mu = \frac{s_m A_{ms} + s_b A_{bs}}{p_m A_{ms} + p_b A_{bs}} \tag{2.7}$$

Let's define:

$$C_1 = \frac{s_m}{p_m}, C_2 = \frac{p_b}{p_m}, C_3 = \frac{s_b}{p_b} \tag{2.8}$$

Therefore, the friction coefficient can be expressed as:

$$\mu = \frac{C_1 A_{ms} + C_2 C_3 A_{bs}}{A_{ms} + C_2 A_{bs}} \tag{2.9}$$

From the above analysis, it can be seen that the contact area of A_{ms} and A_{bs} can be calculated according to the distribution of rough points and the characteristics of the metal contact area and the absorption lubrication film area, so that the friction coefficient can be calculated according to Eq. (2.9). Therefore, it is necessary to obtain the distribution of rough points on the contact surface.

Here, the distribution function of rough points is defined as the ratio of the number of rough points n with interval x to the total number of rough points n_0, which can be expressed by a function of the height of rough points and interval of rough points:

$$\phi = \frac{n}{n_0} = f\left(\frac{x}{H_{\max}}\right) \tag{2.10}$$

where $H_{\max}$ is the height of the rough point.

According to the experimental study of Matlak and Komvopoulos in Ref. [9], the distribution function of rough points is proportional to $x/H_{\max}$. Therefore, the distribution function can be expressed as:

$$\phi = \frac{Dx}{H_{\max}} \tag{2.11}$$

where D is a coefficient of the distribution function of rough points. The number of rough points can be calculated by:

$$n = \frac{n_0 D^2 a_s^2}{H_{\max}^2} \tag{2.12}$$

where a_s represents the critical area of the metal contact and the absorbed lubrication film. According to the contact model proposed by Kragelskii [10], the metal contact area A_{ms} and absorbed lubrication area A_{bs} can be calculated by Eqs. (2.13) and (2.14) respectively:

$$A_{ms} = \frac{\pi R n_0 D^2 a_s^3}{6H_{\max}^2} \tag{2.13}$$

$$A_{bs} = \frac{\pi R n_0 D^2 \{(a_s + t_b)^3 - a_s^3\}}{6H_{\max}^2} \tag{2.14}$$

In the above equation, it is assumed that the end of the rough point is a sphere, the radius is R, and the thickness of the lubrication film is t_b. Substituting Eqs. (2.13) and (2.14) into Eq. (2.5), a cubic equation can be obtained to predict a_s:

$$a_s^3 + 3C_2 t_b a_s^2 + 3C_2 t_b^2 a_s + \left(C_2 t_b^3 - \frac{N}{p_m Q}\right) = 0 \tag{2.15}$$

where

$$Q = \frac{\pi R n_0 D^2}{6H_{\max}^2} \tag{2.16}$$

According to Eqs. (2.8), (2.13) and (2.14), the friction coefficient can be calculated by:

$$\mu = \frac{C_1 a_s^3 + C_2 C_3 \{(a_s + t_b)^3 - a_s^3\}}{a_s^3 + C_2 \{(a_s + t_b)^3 - a_s^3\}} \tag{2.17}$$

where t_b represents the thickness of the lubrication film. When t_b equals to zero, the friction coefficient μ is calculated to be C_1 according to Eq. (2.16). Therefore, the coefficient C_1 represents the friction coefficient under dry cutting condition, which can be obtained according to the dry cutting prediction model. The constants C_2 and C_3 are determined by the lubrication conditions and the performance of the cutting fluid. In this book, C_2 is assumed to be 0.5 under the boundary lubrication condition. The initial value of N in Eq. (2.15) can be obtained by dry cutting condition. The end radius of the rough point R and the height of the rough point $H_{\max}$ can be measured by experiments. The thickness of the lubrication film depends on lubrication conditions. In this book, it is assumed that the thickness of lubrication film t_b is proportional to the flow rate of the cutting fluid, and its proportional coefficient can be determined by the experimental results.

The friction coefficient predicted by the above model is used to predict the cutting force under MQL. The detailed prediction method will be introduced in Chap. 3.

2.3 Cooling Effect of MQL

2.3.1 Cooling Mechanism of MQL

The cooling effect of MQL machining mainly consider two aspects: on one hand, in the cutting process, the air-oil mixture is sprayed into the cutting area under the action of high-pressure high-speed air. When the air-oil mixture particles contact the metal surface with high temperature, the vaporizing center is formed. Air-oil mixture gas liquid two phase boiling rapidly and vaporization take away part of the heat. In the process of cutting, the gas–liquid atomized particles shot from the nozzle to the cutting area constantly impact on the high temperature metal surface, thus forming a forced convection heat transfer, which has a direct cooling effect on the workpiece. On the other hand, due to the lubrication effect of air-oil mixture, the friction force is effectively reduced, resulting in less friction heat. Therefore, it has an indirect cooling effect on the workpiece.

The cooling effect in MQL machining mainly depends on the thermal conductivity of air-oil mixture, such as the thermal conductivity coefficient, specific heat capacity, heat dissipation coefficient, vaporization velocity, and its wettability to metal surface. In addition, the air-oil mixture ratio, and the supply mode of air-oil mixture, including the pressure of the air, the flow rate of cutting fluid, also have a great influence on the cooling effect. The calculation method of heat transfer coefficient in MQL machining is described in details in the following.

2.3.2 Heat Transfer Coefficient in MQL

For the analysis in this book, the air-oil mixture is applied at the tool-workpiece interface through the nozzle under the load of compressed air, as shown in Fig. 2.5. The area between the tool flank surface and the machined surface can be considered as a heat loss area, and the coordinate system is established as shown in Fig. 2.5. In the metal cutting process, the heat dissipation caused by MQL relative to any point on the workpiece can be regarded as a moving heat source, and its heat dissipation density can be calculated according to the following equation:

$$q_{hl} = \overline{h}(T_{flank} - T_w) \tag{2.18}$$

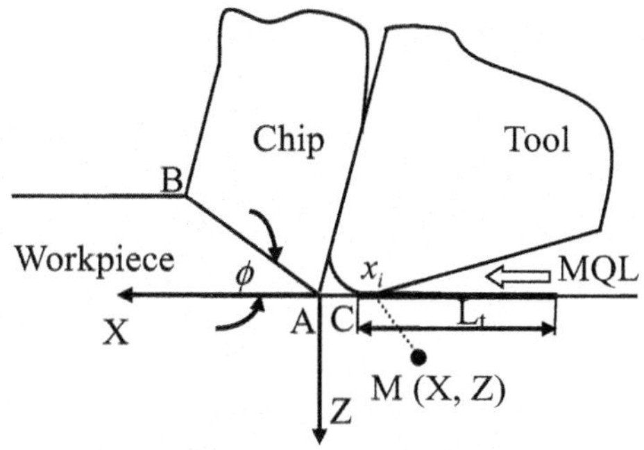

Fig. 2.5 Schematic diagram of heat dissipation in MQL

where $\overline{h}$ is the average convective heat transfer coefficient, T_{flank} is the temperature at tool flank area, and T_w is the ambient temperature, which is assumed to be 20 °C in this book.

As shown in Fig. 2.5, due to the applied air-oil mixture at tool flank area, the air-oil mixture and the machined workpiece can be regarded as forced cooling convection, and the average convective heat transfer coefficient in Eq. (2.18) can be derived from the Nussle number as follows:

$$\overline{N}_u = \frac{\overline{h}L_{eff}}{k_{air}} = 0.664\,\overset{1/3}{\mathrm{Pr}}\,\mathrm{Re}^{1/2} \tag{2.19}$$

where L_{eff} is the effective cooling length, k_{air} is the heat transfer coefficient of air, Pr is the Prandtl number, and Re is the Reynolds number. In MQL cutting process, the air-oil mixture ratio is generally small. In this book, the air-oil mixture ratio is about 1:50. Therefore, the Prandtl number Pr and Reynolds number Re can be estimated by the material properties of the compressed air.

In MQL machining, the air-oil mixture is sprayed to the cutting area under the load of compressed air, so it is necessary to determine whether the compressed air-oil mixture is congested when it passes through the MQL nozzle. In this experiment, the diameter of the nozzle is 0.762 mm. According to the research in Ref. [11], the oil-mist mixture can be judged by the following equation to determine whether congestion phenomenon occurs at the nozzle output:

$$\frac{p^*}{p_{stg}} = \left(\frac{2}{1+\zeta}\right)^{\zeta/(\zeta-1)} \quad (\zeta = 1.4\,\text{For the air}) \tag{2.20}$$

where ξ is the gas heat capacity ratio, and it is also called the adiabatic index. In this analysis, the heat capacity ratio of air is taken to be 1.4. p_{stg} is the stagnant air pressure, p^* is the critical pressure of the air. Equation (2.20) can be used to judge whether the congestion phenomenon of the critical air pressure occurs. If the air pressure at the nozzle outlet is lower than the critical air pressure p^*, the congestion flow phenomenon occurs, and the Mach number is 1. If the congestion phenomenon caused by the air-oil mixture is not considered, the nozzle mouth gas flow pressure, density, temperature and speed can be derived from the following equation:

$$\frac{p}{p_{stg}} = \left\{ \frac{1}{1 + [(\zeta - 1)/2]Ma^2} \right\}^{\zeta/(\zeta-1)} \tag{2.21}$$

$$\frac{\rho}{\rho_{stg}} = \left\{ \frac{1}{1 + [(\zeta - 1)/2]Ma^2} \right\}^{1/(\zeta-1)} \tag{2.22}$$

$$\frac{T}{T_{stg}} = \frac{1}{1 + [(\zeta - 1)/2]Ma^2} \tag{2.23}$$

$$V = Ma\sqrt{RT\zeta} \tag{2.24}$$

where R represents the specific heat capacity of air, Ma is the Mach number.

However, the heat transfer coefficients derived as described above only take into account the cooling effect of air, taking into account that the lubricant will also have a cooling effect due to convection or boiling during the process. Therefore, it is necessary to consider the cooling effect of the air-oil mixture in a comprehensive manner. It is assumed that the cooling effect of the air-oil mixture is proportional to the cooling effect of the air, so the effective heat transfer coefficient under MQL conditions can be expressed by:

$$\overline{h}_{eff} = \lambda \overline{h} \tag{2.25}$$

Since the coefficient λ is used to compensate for the cooling effect of the lubricant, it depends on the air-oil mixture ratio. In the actual experiments, the coefficient λ needs to be calibrated under different air-oil mixture ratios.

The heat loss intensity calculated by Eq. (2.18) can be used to calculate the temperature change due to the cooling effect of MQL, and the details will be introduced in Chap. 3.

2.4 Summary

The effect of MQL on the frictional characteristics of the cutting model is introduced in this chapter, and the lubrication effect and cooling effect of MQL are discussed in details:

(1) The penetration mechanism of cutting medium in the cutting process is introduced. The penetration processes of cutting medium under traditional flood cooling and MQL are compared.
(2) Based on the boundary lubrication theory, the lubrication mechanism and the calculation method of friction coefficient under MQL condition are introduced, its effect on the contact stress during cutting is analyzed.

(3) Based on the forced convection cooling model, the cooling mechanism and the calculation method of the thermal conversion coefficient under MQL condition are introduced, its effect on the thermal stress in the cutting process is analyzed.

References

1. Bian R (2009) Research on the high-speed milling of PH13-8Mo based on cryogenic minimum quantity lubrication. Nanjing University of Aeronautics and Astronautics, Nanjing (in Chinese)
2. Zhang CY (2008) Study on the mechanism of MQL cutting and its application fundament. Jiangsu University, Zhenjiang (in Chinese)
3. Williams J, Tabor D (1977) The role of lubricants in machining. Wear 43(3):275–292
4. Godlevski VA, Volkov AV, Latyshev VN, Maurin LN (1997) The kinetics of lubricant penetration action during machining. Lubr Sci 9(2):127–140
5. Liu JY (2005) Study on action mechanism and experiment with water vapor as coolants and lubricants in green cutting. Harbin Institute of Technology, Harbin (in Chinese)
6. Pereira O, Martin-Alfonso JE, Rodriguez A, Calleja A, Fernandez-Valdivielso A, Lopez de Lacalle LN (2017) Sustainability analysis of lubricant oils for minimum quantity lubrication based on their tribo-rheological performance. J Clean Prod 164:1419–1429
7. Werda S, Duchosal A, Quilliec GL, Leroy R (2019) Effect of minimum quantity lubrication strategies on tribological study of simulated machining operation. Mech Ind 20(6):624
8. Pei HJ, Li F, Chen YY, Wang GC (2022) Atomizing spray characteristics study of minimum quantity lubrication system. Aeronaut Manuf Technol 65(7):70–76 (in Chinese)
9. Matlak J, Komvopoulos K (2015) Friction properties of amorphous carbon ultrathin films deposited by filtered cathodic vacuum arc and radio-frequency sputtering. Thin Solid Films 579:167–173
10. Kragelskii IV (1965) Friction and wear. Butterworths, London
11. Munson BR, Young DF, Okiishi TH (1990) Fundamentals of fluid mechanics. New York

Chapter 3
Force-Temperature Coupled Prediction Model

According to the influence of minimum quantity lubrication (MQL) on the friction characteristics in cutting process, the effect of MQL on the cutting force and cutting temperature are discussed in details. Firstly, the force prediction model in orthogonal cutting is introduced with the given cutting temperature. Secondly, the temperature prediction model is introduced with the given cutting force. Finally, the coupling predictive model of cutting force and cutting temperature under MQL condition is established. Different from other prediction model, the proposed MQL prediction model does not relied upon the measurement of cutting temperature or cutting force, and it can directly predict the cutting force and cutting temperature according to the cutting conditions in the machining process.

3.1 Predictive Modeling of Cutting Force Based on Orthogonal Cutting

Firstly, with the given cutting temperature, the cutting force prediction model for orthogonal cutting is introduced. Two sources of cutting force are considered: (1) the chip formation force due to plastic shear deformation; (2) the plowing force due to the contact between the tool tip and the machined surface.

3.1.1 Chip Formation Force Model

The cutting force model for orthogonal cutting was first proposed by Merchant [1]. Based on the theory of minimum energy, the shear surface is located at the position of minimum shear energy during metal cutting, thus determining the shear angle. Subsequently, Oxley [2] applied the theory of plane plastic strain and proposed the slip line field theory of the principal shear plane, which considers the effects of strain,

© The Author(s), under exclusive license to Springer Nature Singapore Pte Ltd. 2023

X. Ji, *Minimum Quantity Lubrication Machining*,
https://doi.org/10.1007/978-981-19-7087-0_3

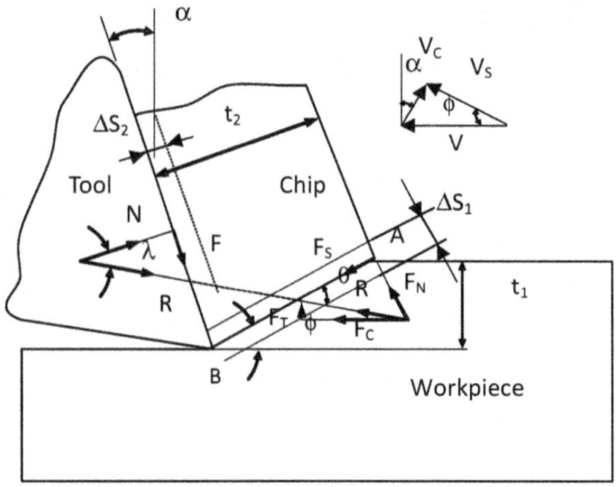

Fig. 3.1 Chip formation force model

strain rate and temperature on flow stress. However, in Oxley's prediction model, the stress is determined by the temperature change rate, and some parameters in the model were obtained by fitting the cutting test data.

Based on Oxley's prediction mode and Johnson–Cook flow stress model, an analytical prediction model of cutting force is established, which is not relied upon the cutting experiments. The variation of material flow stress with strain, strain rate and temperature are considered in the proposed prediction model. The schematic of the prediction model is shown in Fig. 3.1. The basic assumptions of the model are as follows:

(1) The cutting process is at a stable state, and there is no built-up edge. The plane strain state is assumed. The plastic deformation in the cutting layer only occurs at the plane which is perpendicular to the cutting edge.
(2) The first shear zone of the shear plane is assumed to be a narrow strip surrounded by two parallel zones, and the thickness (ΔS_2) of the second shear zone of the cutter-chip interface is assumed to be a constant, which is used to simplify the prediction model.
(3) The temperature and strain on the shear plane are uniformly distributed.
(4) C_{Oxley} and δ are constants of the strain rate, which is used to calculate the strain rates at the shear zone and the tool-chip interface. C_{Oxley} is the ratio of the shear surface length to the thickness of the primary shear zone ΔS_1, which can be expressed as $L_{AB}/\Delta S_1$. δ is the ratio of the thickness of the secondary shear zone ΔS_2 to the thickness of the chip t_2, which can be denoted as$\Delta S_2/t_2$.

As shown in Fig. 3.1, the stress distribution at the shear surface AB and tool-chip interface is analyzed in the chip formation force model. When the shear angle ϕ is determined, the cutting force can be calculated by the equilibrium equation of the shear plane and tool-chip interface. The chip thickness t_2 and other cutting force components can be calculated by Eq. (3.1).

$$
\begin{aligned}
t_2 &= t_1 \cos(\phi - \alpha) / \sin \phi \\
F_c &= R \cos(\lambda - \alpha) \\
F_t &= R \sin(\lambda - \alpha) \\
F &= R \sin \lambda \\
N &= R \cos \lambda \\
R &= \frac{F_S}{\cos \theta} = \frac{k_{AB} t_1 w}{\sin \phi \cos \theta}
\end{aligned}
\tag{3.1}
$$

The Johnson–Cook model is used to predict the flow stress, because a large plastic deformation process with high temperature, high pressure and high strain rate occurs during the cutting process. The flow stress is established as a function of strain, strain rate and temperature [3]. The general expression of Johnson–Cook flow stress can be expressed by Eq. (3.2):

$$
\sigma = \left(A + B\varepsilon_p^n\right)\left(1 + C \ln \frac{\dot{\varepsilon}_p}{\dot{\varepsilon}_0}\right)\left(1 - \left(\frac{T - T_w}{T_m - T_w}\right)^m\right) \tag{3.2}
$$

where σ is the effective stress, ε_p is the effective plastic strain, $\dot{\varepsilon}_p$ is the effective plastic strain rate, $\dot{\varepsilon}_0$ is the reference strain rate, and sometimes it is assumed to be $1\ s^{-1}$. T is the current temperature value, T_m is the melting point of the material, T_w is the ambient temperature, A, B, C, m, n are Johnson–Cook flow stress constants respectively.

The stress, strain, and temperature distributions at the shear plane and the tool-chip interface are analyzed as follows:

(1) Shear plane

The chip velocity and shear surface flow velocity can be calculated from Fig. 3.1, and they can be expressed by Eqs. (3.3) and (3.4):

$$
V_c = \frac{\sin \phi}{\cos(\phi - \alpha)} V \tag{3.3}
$$

$$
V_s = \frac{\cos \alpha}{\cos(\phi - \alpha)} V \tag{3.4}
$$

where V refers to the cutting speed, V_c refers to the chip speed, V_s refers to the flow speed of the shear plane. α is the tool rake angle, ϕ is the shear angle.

For any particle in the shear zone, the velocity increment and displacement increment of this particle in the X and Y directions can be expressed by Eq. (3.5):

$$
\begin{aligned}
&\Delta v_x = \frac{-V \cos \alpha \cos \phi}{\cos(\phi - \alpha)}, \ \Delta v_y = \frac{V \cos \alpha \sin \phi}{\cos(\phi - \alpha)} \\
&\Delta x = \frac{\Delta S_1}{\sin \phi}, \ \Delta y = \frac{\Delta S_1}{\cos \phi}
\end{aligned}
\tag{3.5}
$$

The average shear strain rate at the shear zone can be expressed by the flow velocity as follows:

$$\dot{\gamma}_{avg} = \sqrt{\left(\frac{\partial v_x}{\partial y} + \frac{\partial v_y}{\partial x}\right)^2 + 4\left(\frac{\partial v_x}{\partial x}\right)^2} \tag{3.6}$$

Substituting Eqs. (3.5) into (3.5), it can be obtained:

$$\dot{\gamma}_{avg} = \frac{V\cos\alpha}{\Delta S_1\cos(\phi - \alpha)} \tag{3.7}$$

According to Von Mises stress yield criterion, the equivalent strain rate and equivalent strain of shear plane AB can be expressed as:

$$\dot{\varepsilon}_{AB} = \frac{1}{\sqrt{3}}\frac{V\cos\alpha}{\Delta S_1\cos(\phi - \alpha)} \tag{3.8}$$

$$\varepsilon_{AB} = \frac{1}{2\sqrt{3}}\frac{\cos\alpha}{\sin\phi\cos(\phi - \alpha)} \tag{3.9}$$

According to the Johnson–Cook flow stress model, the flow stress at shear plane can be calculated by Eq. (3.10):

$$k_{AB} = \frac{1}{\sqrt{3}}\left(A + B\varepsilon_{AB}{}^{n}\right)\left(1 + C\ln\frac{\dot{\varepsilon}_{AB}}{\dot{\varepsilon}_0}\right)\left(1 - \left(\frac{T_{AB} - T_w}{T_m - T_w}\right)^m\right) \tag{3.10}$$

where k_{AB} is the flow stress at the shear plane, ε_{AB} is the effective plastic strain of the shear plane, $\dot{\varepsilon}_{AB}$ is the effective plastic strain rate of the shear plane, and $\dot{\varepsilon}_0$ is the reference strain rate, and it is assumed to be 1 s^{-1} in this analysis. T_{AB} is the temperature at the shear plane, T_m is the melting point of the material, and T_w is the ambient temperature, which is assumed to be 20 °C in this paper. *A*, *B*, *C*, *m* and *n* represent Johnson–Cook flow stress constants respectively.

Once the flow stress at the shear plane is obtained, the remaining cutting force components can be calculated by Eq. (3.1). The friction angle can be calculated by Eq. (3.11):

$$\lambda = \theta + \alpha - \phi \tag{3.11}$$

The friction coefficient can be calculated according to the angle friction. According to Oxley's prediction model, the included angle θ between the resultant force component *R* and the shear plane AB in Fig. 3.1 can be calculated:

$$\tan\theta = 1 + 2\left(\frac{\pi}{4} - \phi\right) - C_n \tag{3.12}$$

where C_n is different from the definition of Oxley's prediction model. In this paper, the influence of material strain on the flow stress is considered, which can be calculated by Eq. (3.13):

$$C_n = C_{Oxley} n \frac{B\varepsilon^n_{AB}}{A + B\varepsilon^n_{AB}} \tag{3.13}$$

The average temperature at the shear plane AB can be obtained by Eq. (3.14):

$$T_{AB} = T_w + \eta \Delta T_{sz} \tag{3.14}$$

where η is the percentage of total shear energy converted to enthalpy, which is set to be 0.9 in this book. $(1 - \eta)$ is the potential energy stored in deformed chip. ΔT_{sz} is the temperature rise caused by the plastic strain of the material at primary deformation zone, which generated the shear energy $F_s V_s$. m_{chip} is the chip mass per unit time, and it can be calculated as $m_{chip} = \rho_{wk} V t_1 w$. Therefore, ΔT_{sz} can be calculated by Eq. (3.15):

$$\Delta T_{sz} = \frac{(1 - \beta) F_s V_s}{m_{chip} C_p} \tag{3.15}$$

where C_p is the specific heat capacity of the workpiece material, which can be expressed as a function of cutting temperature. β is the heat distribution coefficient at the shear zone, which can be calculated by Eq. (3.16):

$$\begin{aligned} \beta &= 0.5 - 0.35 \log_{10}(R_T \tan\varphi) \; for \; 0.004 \le R_T \tan\varphi \le 10 \\ \beta &= 0.3 - 0.15 \log_{10}(R_T \tan\varphi) \; for \; R_T \tan\varphi > 10 \end{aligned} \tag{3.16}$$

where R_T is a dimensionless coefficient, which can be calculated by Eq. (3.17):

$$R_T = \frac{\rho_{wk} C_p V t_1}{K_{wk}} \tag{3.17}$$

where ρ_{wk} is the density of the workpiece, K_{wk} is the thermal conductivity of the workpiece, which can be expressed as a function of the cutting temperature.

Form Fig. 3.1, it is also found that the normal stress of point B can be calculated by combining the boundary stress with Johnson–Cook flow stress model:

$$\sigma'_N = k_{AB}\left(1 + \frac{\pi}{2} - 2\alpha - 2C_n\right) \tag{3.18}$$

(2) Tool and chip interface

The plastic region of the tool-chip interface can be assumed to be a rectangular region, and its thickness ΔS_1 can be assumed to be a constant, $\Delta S_1 = \delta t_2$. Therefore, the effective plastic strain rate at tool-chip interface can be calculated by Eq. (3.19):

$$\dot{\varepsilon}_{int} = \frac{1}{\sqrt{3}} \frac{V_C}{\delta t_2} \tag{3.19}$$

The contact length of tool-chip can be calculated according to the torque balance formula of shear surface:

$$h = \frac{t_1 \sin\theta}{\cos\lambda \sin\phi}\left\{1 + \frac{C_n}{3\tan\theta}\right\} \tag{3.20}$$

Assuming that the stress at the tool-chip interface is uniformly distributed, and it can be calculated by Eq. (3.21):

$$\tau_{int} = \frac{F}{hw} \tag{3.21}$$

According to the stress analysis of the tool-chip interface, the stress at point B can be calculated by Eq. (3.22):

$$\sigma_N = \frac{N}{hw} \tag{3.22}$$

The average temperature of tool-chip interface can be expressed as:

$$T_{\text{int}} = T_w + \Delta T_{sz} + \Psi \Delta T_M \tag{3.23}$$

where ΔT_M is the maximum temperature rise of tool-chip interface. Ψ is the calibration coefficient, which is used to considering the average temperature at tool-chip interface. Assuming that the heat source at the tool-chip interface is a rectangular region, ΔT_M can be calculated by Eq. (3.24) [4]:

$$\log_{10}\left(\frac{\Delta T_M}{\Delta T_C}\right) = 0.006 - 0.195(\delta)\sqrt{\frac{R_T t_2}{t_1}} + 0.5\log_{10}\left(\frac{R_T t_2}{h}\right) \tag{3.24}$$

where ΔT_C is the average temperature rise of chips, which can be calculated by Eq. (3.25):

$$\Delta T_C = \frac{F V_C}{m_{chip} C_p} \tag{3.25}$$

In summary, the average flow stress of the chip can be expressed as:

$$k_{chip} = \frac{1}{\sqrt{3}}\left(A + B\varepsilon_{int}^n\right)\left(1 + C\ln\frac{\dot{\varepsilon}_{int}}{\dot{\varepsilon}_0}\right)\left(1 - \left(\frac{T_{int} - T_w}{T_m - T_w}\right)^m\right) \tag{3.26}$$

where the effective strain ε_{int} in the chip can be obtained by Eq. (3.27):

$$\varepsilon_{int} = 2\varepsilon_{AB} + \frac{1}{\sqrt{3}}\frac{h}{\delta t_2} \tag{3.27}$$

In the cutting model, the stress τ_{int} at the tool-chip interface and the flow stress k_{chip} in the chip can be calculated for each shear angle increment ϕ, and the shear angle is the maximum value when the stress at the tool-chip interface is closest to the stress in the chip. The strain rate coefficient C_{Oxley} in the primary plastic deformation zone can be determined according to the position where the normal stress at the tool-chip interface is closest to the boundary stress of point B. The strain rate coefficient δ of the secondary plastic deformation zone is determined according to the principle of minimum cutting force. The flowchart of chip formation prediction model for orthogonal cutting is shown in Fig. 3.2.

The output variables of the prediction model include the shear angle ϕ, the flow stress at the shear plane k_{AB}, the cutting force along the cutting direction F_c and the thrust force perpendicular to the machined surface F_t. The prediction model is only applicable to dry cutting condition, and the friction angle is calculated according to the force balance and material properties. However, for machining with cutting fluid, if the lubrication coefficient is known, the friction angle can be calculated by the lubrication coefficient:

$$\lambda = \arctan(\mu) \tag{3.28}$$

For MQL machining, the friction coefficient can be predicted by the boundary lubrication model introduced in Chap. 2. The results calculated by Eq. (2.17) can be substituted into Eq. (3.28) to calculate the friction angle under MQL condition, so as to predict the chip formation force.

3.1.2 Plowing Force Model

The chip formation force prediction model assumes that the tool tip is completely sharp, but actually every tool has a tool edge radius. For ordinary machining, the effect of the tool edge radius can be ignored. However, for precision cutting, the influence of tool edge radius must be considered when the cutting depth and tool tip radius are in the same order of magnitude. The force produced by the tool tip contacting the machined surface is called plowing force. A lot of researches were focused on the plowing force model [5–11].

The plowing force model was firstly proposed by Waldorf [12]. Based on the slip line theory, Waldorf put forward the prediction model of plowing generated by the tool edge radius, as shown in Fig. 3.3.

In Fig. 3.3, r_e is the cutting-edge radius, α is the rake angle of the tool, ϕ is the shear angle, and t_1 is the thickness of undeformed chip. The angles θ_{fan}, γ_{plow} and

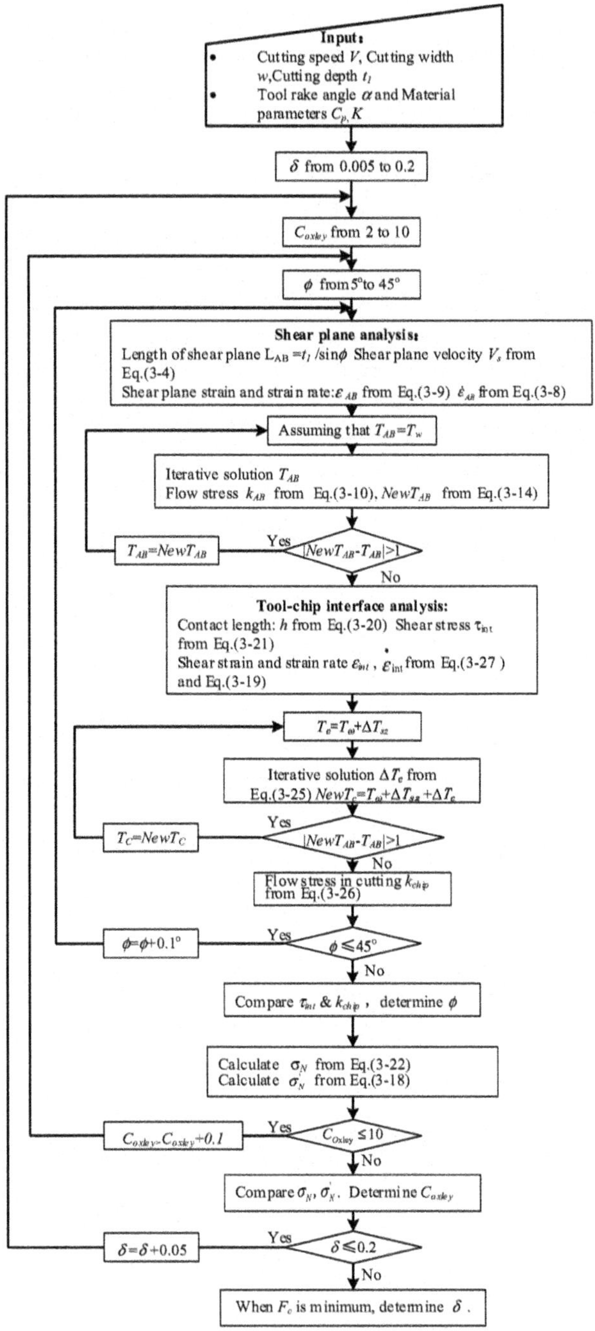

Fig. 3.2 Flow chart of chip formation prediction model

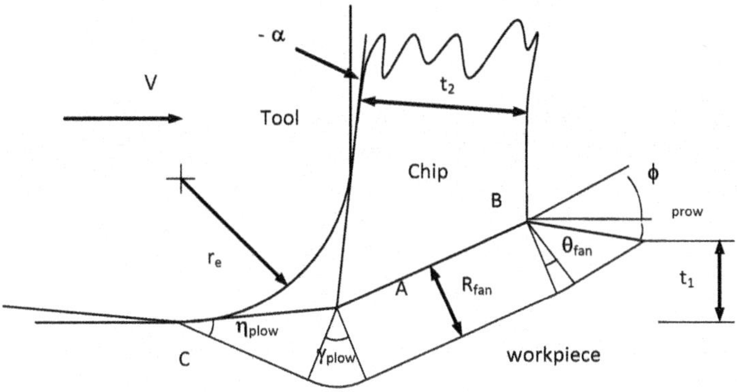

Fig. 3.3 Waldorf's slip line field for plowing [12]

η_{plow} of the sector in the figure can be calculated according to geometric and friction relations according to the following Eq. (3.29):

$$\begin{aligned} \theta_{fan} &= \frac{\pi}{4} - \rho_{prow} - \phi \\ \gamma_{plow} &= \eta_{plow} + \phi - \sin^{-1}\left(\sqrt{2}\sin(\rho_{prow})\sin(\eta_{plow})\right) \\ \eta_{plow} &= 0.5 \cdot \cos^{-1}(\mu_{plow}) \end{aligned} \tag{3.29}$$

In Eq. (3.29), ρ_{prow} is the included angle between the unprocessed convex part and the horizontal plane caused by the cutting-edge radius, as shown in Fig. 3.3. μ_{plow} is the friction factor, which is assumed to be 0.99 in this paper. This is a reasonable assumption because the shear stress at the point of contact between the workpiece material and the chip tumor should be close to the chip flow stress. R_{fan} is the sector radius of the sector, which can be obtained by Eq. (3.30):

$$R_{fan} = \sqrt{\left(r_e \cdot \tan\left(\frac{\pi}{4} + \frac{\alpha}{2}\right) + \frac{\sqrt{2} \cdot R_{fan} \cdot \sin(\rho_{prow})}{\tan\left(\frac{\pi}{2} + \alpha\right)}\right)^2 + 2 \cdot \left[R_{fan} \cdot \sin(\rho_{prow})\right]^2 \cdot \sin(\eta_{plow})} \tag{3.30}$$

If the shear angle ϕ and the flow stress k_{AB} on the shear plane are given, the plowing force can be calculated by Eq. (3.31). Among them, P_{cut} is the plowing force in the cutting direction, P_{thrust} is the plowing force perpendicular to the direction of the machined surface, and w is the cutting width.

$$\begin{aligned} P_{cut} &= k_{AB} \cdot w \begin{bmatrix} \cos(2\eta_{plow})\cos(\phi - \gamma_{plow} + \eta_{plow}) + \\ (1 + 2\theta_{fan} + 2\gamma_{plow} + \sin(2\eta_{plow}))\sin(\phi - \gamma_{plow} + \eta_{plow}) \end{bmatrix} \cdot CA \\ P_{thrust} &= k_{AB} \cdot w \begin{bmatrix} (1 + 2\theta_{fan} + 2\gamma_{plow} + \sin(2\eta_{plow}))\cos(\phi - \gamma_{plow} + \eta_{plow}) - \\ \cos(2\eta_{plow})\sin(\phi - \gamma_{plow} + \eta_{plow}) \end{bmatrix} \cdot CA \end{aligned} \tag{3.31}$$

where

$$CA = \frac{R_{fan}}{\sin \eta_{plow}} \tag{3.32}$$

In Waldorf's plowing force model, the ρ_{prow} angle depends on the radius of the tool edge radius. When the tool edge radius is large, ρ_{prow} is 0° and the predicted results are in good agreement with the measured results. When the tool edge radius is small, ρ_{prow} tends to be larger. In this book, ρ_{prow} is taken as 10°, which is within a reasonable range according to Ref. [12].

$$\begin{aligned} F_C &= F_c + P_{cut} \\ F_T &= F_t + P_{thrust} \end{aligned} \tag{3.33}$$

3.2 Predictive Modeling of Cutting Temperature

In this book, the temperature prediction model is developed based on the temperature model which is proposed by Komanduri. Four heat sources/losses are considered in this book as shown in Fig. 3.4: (1) shear heat source due to shear deformation; (2) primary friction heat source at tool-chip interface; (3) secondary friction heat source at tool-workpiece interface; (4) heat loss due to minimum quantity lubrication (MQL). The moving and fixed heat source method proposed by Jaeger is applied in the proposed model. The temperature prediction modeling of the workpiece, chip and tool are presented separately.

3.2.1 Temperature Model in Workpiece

The temperature model in workpiece under MQL condition mainly considers three heat sources: (1) primary heat source due to shear deformation; (2) rubbing heat source due to friction between the tool and the machined surface; (3) heat loss due to

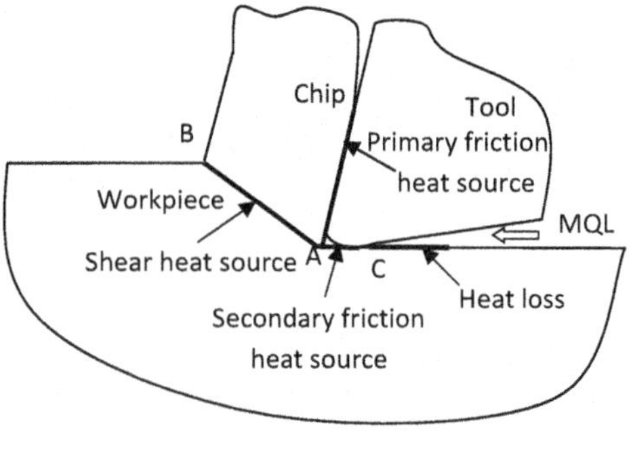

Fig. 3.4 Heat sources/losses in MQL machining

MQL. For any point in the workpiece, all heat sources can be regarded as moving heat sources. Assuming that the unmachined workpiece surface is an adiabatic boundary, the imaginary heat source method is used to calculate the temperature rise generated by the heat source. The temperature changes of workpiece caused by these three heat sources are introduced respectively.

From Fig. 3.5, the temperature rises of point M (X, Z) in the workpiece due to shear plastic deformation, which is caused by the combination of shear heat source and its imaginary heat source. The temperature change of point M (X, Z) can be calculated by Eq. (3.34):

$$\Delta T_{wk\text{-}shear}(X, Z) = \frac{q_{shear}}{2\pi k_{wk}} \int_0^{L_{AB}} e^{-\frac{(X - l_i \sin\varphi)V}{2a_{wk}}} \left\{ K_0\left[\frac{V}{2a_{wk}}\sqrt{(X - l_i\cos\phi)^2 + (Z + l_i\sin\phi)^2}\right] + K_0\left[\frac{V}{2a_{wk}}\sqrt{(X - l_i\cos\phi)^2 + (2t_1 - l_i\sin\phi + Z)^2}\right]\right\} dl_i \tag{3.34}$$

where L_{AB} is the length of shear surface, which equals to $\frac{t_1}{\sin\phi}$. q_{shear} is the heat source density of shear plane, V is the cutting speed, k_{wk} and a_{wk} is the thermal conductivity and thermal diffusion coefficient of the workpiece, respectively. And t_1 is the undeformed chip thickness, ϕ is the shear angle, K_0 is the modified Bessel function of the second kind.

From Fig. 3.6, the secondary friction heat source between the tool tip and machined surface can be regarded as a moving heat source along the X axis. The temperature change caused by the secondary friction heat source can be calculated by Eq. (3.35) [13, 14].

$$\Delta T_{wk\text{-}rub}(X, Z) = \frac{q_{rub}}{\pi k_{wk}} \int_0^{CA} \gamma e^{-\frac{(-X - x_i)V}{2a_{wk}}} \left\{ K_0\left(\frac{V}{2a_{wk}}\sqrt{(X + x_i)^2 + Z^2}\right)\right\} dx_i \tag{3.35}$$

where q_{rub} is the heat source density of the secondary friction heat source at the tool-workpiece interface. CA can be calculated by Eq. (3.32). The heat distribution coefficient of the secondary friction heat source γ can be calculated by Eq. (3.36), which is dependent on the characteristics of the tool and the workpiece. In Eq. (3.36),

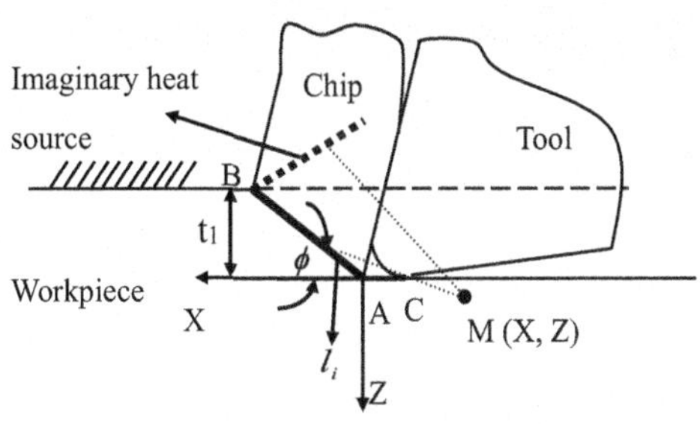

Fig. 3.5 Schematic of the primary heat source for workpiece

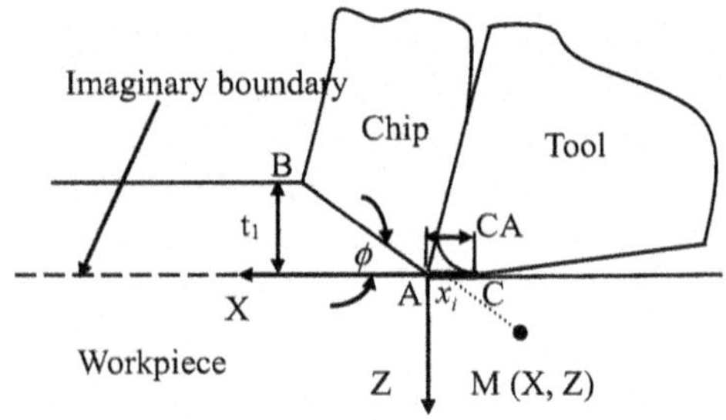

Fig. 3.6 Schematic of the rubbing heat source for workpiece

k_{wk}, ρ_{wk}, C_p, k_t, ρ_t and C_t are the thermal conductivity, density and specific heat capacity of the workpiece and the tool, respectively.

$$\gamma = \frac{\sqrt{k_{wk}\rho_{wk}C_p}}{\sqrt{k_{wk}\rho_{wk}C_p} + \sqrt{k_t\rho_t C_t}} \tag{3.36}$$

From Fig. 3.7, it is found that the minimum quantity lubrication is applied at the interface between the tool flank face and the machined workpiece surface. Therefore, the heat loss due to MQL can be considered as a moving heat loss along the X-axis. The temperature change caused by MQL is calculated by Eq. (3.37).

$$\Delta T_{wk\text{-}hl}(X, Z) = \frac{q_{hl}}{\pi k_{wk}} \int_0^{Lt} e^{-\frac{(-X+L_{AB}\cos\phi+x_i)V}{2a_{wk}}} \left\{ K_0\left(\frac{V}{2a_{wk}} \sqrt{(X - L_{AB}\cos\phi - x_i)^2 + (Z+t_1)^2} \right) \right\} dx_i \tag{3.37}$$

where L_t is the effective cooling length, q_{hl} is the heat loss density due to MQL, which is calculated by Eq. (2.18) in Chap. 2.

The heat source density in Eqs. (3.34) and (3.35) can be calculated according to the cutting forces and cutting parameters:

$$q_{shear} = \frac{(F_c\cos\phi - F_t\sin\phi)(V\cos\alpha / \cos(\phi - \alpha))}{t_1 \cdot w \cdot \csc\phi} \tag{3.38}$$

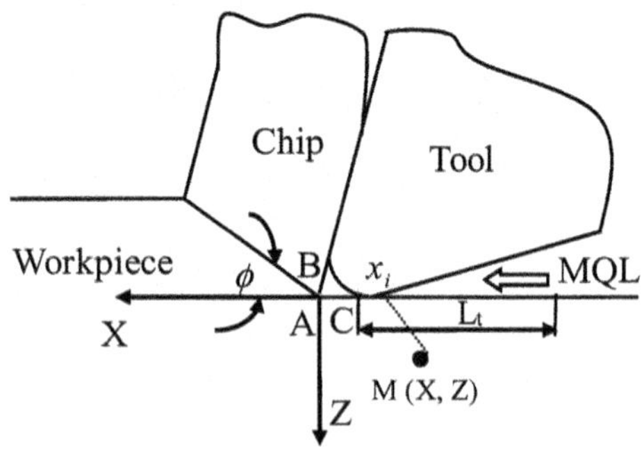

Fig. 3.7 Schematic of the heat loss for workpiece in MQL machining

$$q_{rub} = \frac{P_{cut} V}{w \cdot CA} \tag{3.39}$$

To sum up, the temperature change at any point of workpiece in MQL machining can be calculated by superposition of the above heat sources and heat loss:

$$\Delta T_{total}(X, Z) = \Delta T_{wk\text{-}shear}(X, Z) + \Delta T_{wk\text{-}rub}(X, Z) - \Delta T_{wk\text{-}hl}(X, Z) \tag{3.40}$$

3.2.2 Temperature Model in Chip

In MQL machining, the temperature change of chip mainly considers two heat sources: (1) primary heat source due to shear deformation; (2) secondary heat source due to friction at tool-chip interface. For any point in the chip, all heat sources can be regarded as moving heat sources. Assuming that the back of chip is an adiabatic boundary, the imaginary heat source method is used to calculate the temperature change due to heat source.

From Fig. 3.8, it is found that the temperature change of point M (X, Z) in the chip due to shear plastic deformation can be regarded as a combination of shear heat source and its imaginary heat source. The temperature change of point M (X, Z) can be calculated by Eq. (3.41):

$$\Delta T_{ch-s} = \frac{q_{shear}}{2\pi k_{wk}} \int_{l_i=0}^{L_{AB}} e^{\frac{-(X-x_i)V_c}{2a_{wk}}} \left\{ \begin{array}{l} K_0\left(\frac{V_c}{2a_{ch}}\sqrt{(X-x_i)^2+(Z+z_i)^2}\right) \\ +K_0\left(\frac{V_c}{2a_{ch}}\sqrt{(X-x_i)^2+((2t_2-z_i)+Z)^2}\right) \end{array} \right\} dl_i \tag{3.41}$$

where $x_i = h - l_i \sin(\phi - \alpha)$, $z_i = l_i \cos(\phi - \alpha)$, the length of shear plane $L_{AB} = \frac{t_1}{\sin\phi}$, h is the contact length of the tool-chip interface, which can be calculated by Eq. (3.20). ϕ is the shear angle, α is the rake angle of tool, t_1 is the undeformed chip thickness.

From Fig. 3.9, it can be seen that the primary friction heat source at the tool-chip interface can be regarded as a moving heat source for any point in the chip, and the

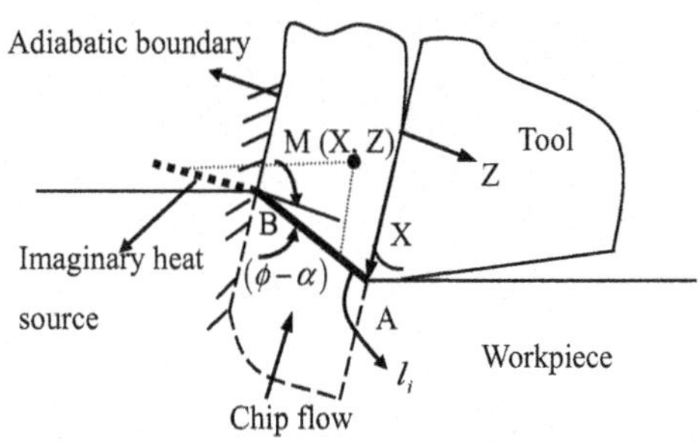

Fig. 3.8 Schematic of the primary heat source for chip

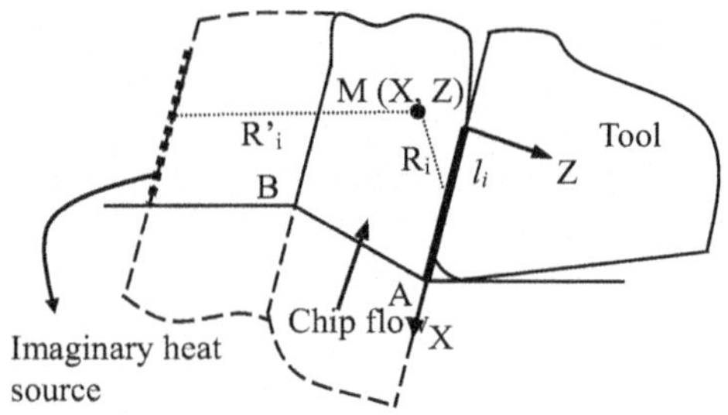

Fig. 3.9 Schematic of the secondary heat source for chip

temperature change of chip can be calculated by Eq. (3.42).

$$\Delta T_{ch-f} = \frac{q_f}{\pi k_{wk}} \int_{l_i=0}^{h} B_1(x) e^{-\frac{(X-l_i)V_c}{2a_{ch}}} \left\{ 2K_0\left(\frac{R_i V_c}{2a_{ch}}\right) + 2K_0\left(\frac{R_i' V_c}{2a_{ch}}\right) \right\} dl_i \tag{3.42}$$

where $R_i = \sqrt{(X - l_i)^2 + Z^2}$, $R_i' = \sqrt{(X - l_i)^2 + (Z + 2t_2)^2}$. $B_1(x)$ is the distribution coefficient of friction heat source, which represents the percentage of heat entering into the chip. The total temperature rise in the chip caused by shear heat source and primary friction heat source is $\Delta T_{ch-s} + \Delta T_{ch-f}$.

According to the predicted cutting force and the tool-chip contact length, the primary friction heat source density q_f can be calculated by Eq. (3.43).

$$q_f = \frac{F V_c}{wh} \tag{3.43}$$

where F is the friction force at tool-chip interface, V_c is the flow velocity of the chip, w is the cutting width, h is the tool-chip contact length, which can be calculated by Eq. (3.20).

3.2.3 Temperature Model in Tool

In MQL machining, the temperature change in the tool mainly considers three heat sources: (1) the primary friction heat source at tool-chip interface; (2) the secondary friction heat source at tool-workpiece interface; (3) heat loss between tool flank and the machined workpiece surface. For any point in the tool, all heat sources/loss can be regarded as moving heat sources. Assuming that the unmachined surface workpiece is an adiabatic boundary, the imaginary heat source method is applied to calculate the temperature change in the tool.

The coordinate system of tool is established as shown in Fig. 3.10. The temperature change of point M (X, Z) in the tool due to tool-chip friction, which is caused by the

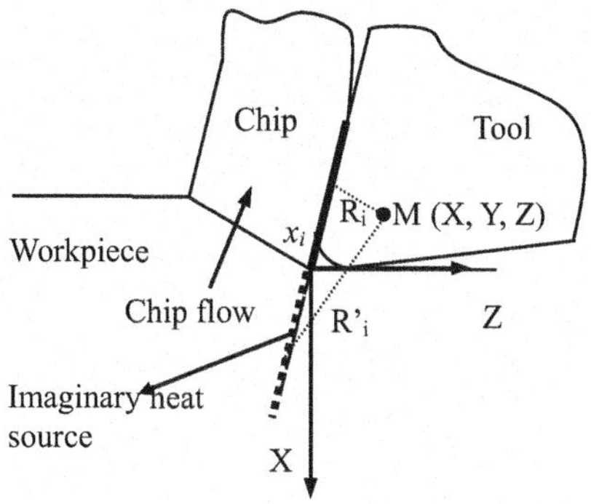

Fig. 3.10 Schematic of the secondary heat source for tool

combination of primary friction heat source and its imaginary heat sources can be calculated by Eq. (3.44):

$$\Delta T_{t-f} = \frac{q_f}{2\pi k_t}\int_0^h (1 - B_1(x_i))\int_{-w/2}^{w/2}\left(\frac{1}{R_i} + \frac{1}{R_i'}\right)dy_i dx_i \tag{3.44}$$

where
$R_i = \sqrt{(X + x_i\cos(\alpha + \beta_{clearance}))^2 + (Y - y_i)^2 + (Z - x_i\cos(\alpha + \beta_{clearance}))^2}$,
$R_i' = \sqrt{(x_i\cos(\alpha + \beta_{clearance}) - X)^2 + (Y - y_i)^2 + (Z + x_i\sin(\alpha + \beta_{clearance}))^2}$.
α is the tool rake angle, $\beta_{clearance}$ is the tool clearance angle, and w is the cutting width. $1 - B_1(x_i)$ represents the percentage of primary friction heat entering into the tool. q_f is the primary friction heat source density at tool-chip interface, which can be calculated by Eq. (3.43).

As shown in Fig. 3.11, the temperature change of point M (X, Z) in the tool due to secondary friction heat source at tool-workpiece interface can be calculated by Eq. (3.45):

$$\Delta T_{t-rub} = \frac{q_{rub}}{2\pi k_t}\int_0^{CA}(1-\gamma)\int_{-w/2}^{w/2}\left(\frac{1}{R_i} + \frac{1}{R_i'}\right)dy_i dx_i \tag{3.45}$$

where $R_i = \sqrt{X^2 + (Y - y_i)^2 + (Z - x_i)^2}$, $R_i' = \sqrt{X^2 + (Y - y_i)^2 + (Z + x_i)^2}$. $1 - \gamma$ represents the percentage of secondary friction heat entering into the tool. CA can be calculated by the Eq. (3.32). q_{rub} is the secondary friction heat source density at tool-workpiece interface, which can be calculated by Eq. (3.39).

In MQL machining, the air-oil mixture is applied between the tool flank and the machined surface, as shown in Fig. 3.12. Therefore, it is assumed that there is a heat loss at tool-workpiece interface. The cutting-edge length can be considered as the length of the heat loss region, and the cutting width can be considered as the width of the heat loss. From Fig. 3.12, for any point in the tool, the heat loss can be regarded as

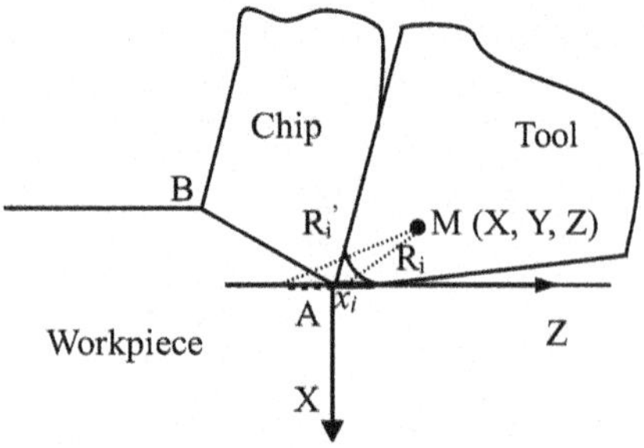

Fig. 3.11 Schematic of the rubbing heat source for tool

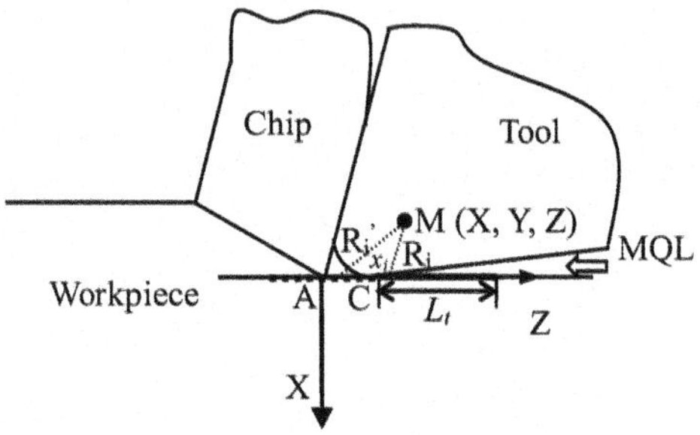

Fig. 3.12 Schematic of the heat loss for tool due to MQL

a fixed heat loss. The temperature change of the tool caused by MQL can be regarded as the superposition of heat loss and its imaginary heat loss, which can be calculated by Eq. (3.46).

$$\Delta T_{t-hl} = \frac{q_{hl}}{2\pi k_t} \int_{0}^{L_t} \int_{-w/2}^{w/2} \left(\frac{1}{R_i} + \frac{1}{R_i'} \right) dy_i dx_i \tag{3.46}$$

where $R_i = \sqrt{X^2 + (Y - y_i)^2 + (Z - CA - x_i)^2}$,$R_i' = \sqrt{X^2 + (Y - y_i)^2 + (Z - CA + x_i)^2}$. q_{hl} is the heat loss density and can be calculated by Eq. (2.18).

In summary, the temperature variation at any point in the tool can be expressed by $\Delta T_{t-f} + \Delta T_{t-rub} - \Delta T_{t-hl}$.

Assuming that the temperature in the tool and the temperature in the chip is the same at tool-chip interface. Therefore, the following relationship exists at the rake face of the tool:

$$\Delta T_{ch-s} + \Delta T_{ch-f} + T_0 = \Delta T_{t-f} + \Delta T_{t-rub} - \Delta T_{t-hl} + T_0 \tag{3.47}$$

Assuming that there are n points at the tool-chip contact interface, the heat distribution coefficient $B_1(x_i)$ of the Eqs. (3.42) and (3.42) can be obtained by the n equations established by Eq. (3.47).

3.3 Force-Temperature Coupled Model

The above prediction models require the measured cutting force or cutting temperature as inputs. In this section, a coupled predictive modelling of cutting temperature and cutting force in MQL is presented. The prediction model is developed based on the modified Oxley's prediction theory, considering both the lubrication and cooling effects due to MQL. The cutting force and cutting temperature can be directly predicted according to the cutting parameters, tool geometry, material properties and MQL parameters, which is not relied upon the measured cutting force or cutting temperature.

3.3.1 *Modified Oxley's Predictive Model*

The Oxley's prediction model is established for dry cutting without considering the effect of cutting fluid, and the friction angle is calculated by the force balance equation and material properties, as shown in Eq. (3.11). However, in MQL machining, the air-oil mixture has both lubrication and cooling effects on the cutting process. Therefore, the Oxley's predictive model need to be modified from the lubrication and cooling aspects.

For the lubrication effect of MQL, the friction coefficient in MQL can be calculated by Eq. (2.17). The friction angle can be obtained by the friction coefficient, and the predicted friction angle can be substituted into Eq. (3.11).

For the cooling effect of MQL, according to the predictive model of workpiece temperature introduced in Sect. 3.2.1, the average temperature at shear plane and tool-chip interface can be predicted by Eqs. (3.48) and (3.49).

$$T_{AB} = \frac{q_{shear}}{2\pi k_{wk}} \int_0^{L_{AB}\cos\phi} \int_0^{L_{AB}} e^{-\frac{(X-l_i\sin\varphi)V}{2a_{wk}}} \left\{ K_0\left[\frac{V}{2a_{wk}}\sqrt{(x_i - l_i\cos\phi)^2 + (l_i\sin\phi)^2}\right] + K_0\left[\frac{V}{2a_{wk}}\sqrt{(x_i - l_i\cos\phi)^2 + (2t_1 - l_i\sin\phi)^2}\right]\right\} dx_i dl_i \tag{3.48}$$

$$T_{\text{int}} = \frac{q_f}{\pi k_{ch}} \int_0^h \int_{l_i=0}^h B_1(x) e^{-\frac{(X-l_i)V_c}{2a_{ch}}} \left\{ 2K_0\left(\frac{R_i V_c}{2a_{ch}}\right) + 2K_0\left(\frac{R_i' V_c}{2a_{ch}}\right)\right\} dx_i dl_i \tag{3.49}$$

where $R_i = \sqrt{(x_i - l_i)^2}$, $R_i' = \sqrt{(X - l_i)^2 + (2t_2)^2}$. L_{AB} is the length of the shear plane, and $L_{AB} = t_1/\sin\phi$ is the contact length of the tool-chip interface, which is calculated by Eq. (3.20).

Two kinds of modified Oxley's prediction models are defined:

(1) The modified Oxley's prediction model I, which only considers the lubrication effect of MQL. The calculation flowchart is shown in Fig. 3.13.
(2) The modified Oxley's prediction model II, which considers the lubrication and cooling effects of MQL. The calculation flowchart is shown in Fig. 3.14.

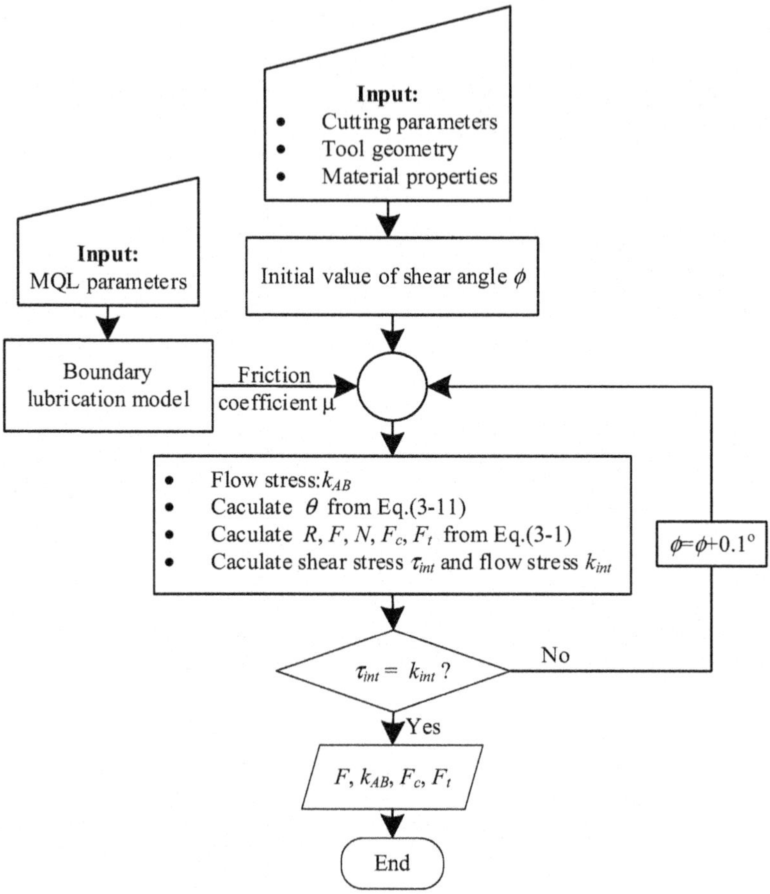

Fig. 3.13 Schematic of the modified Oxley's model I

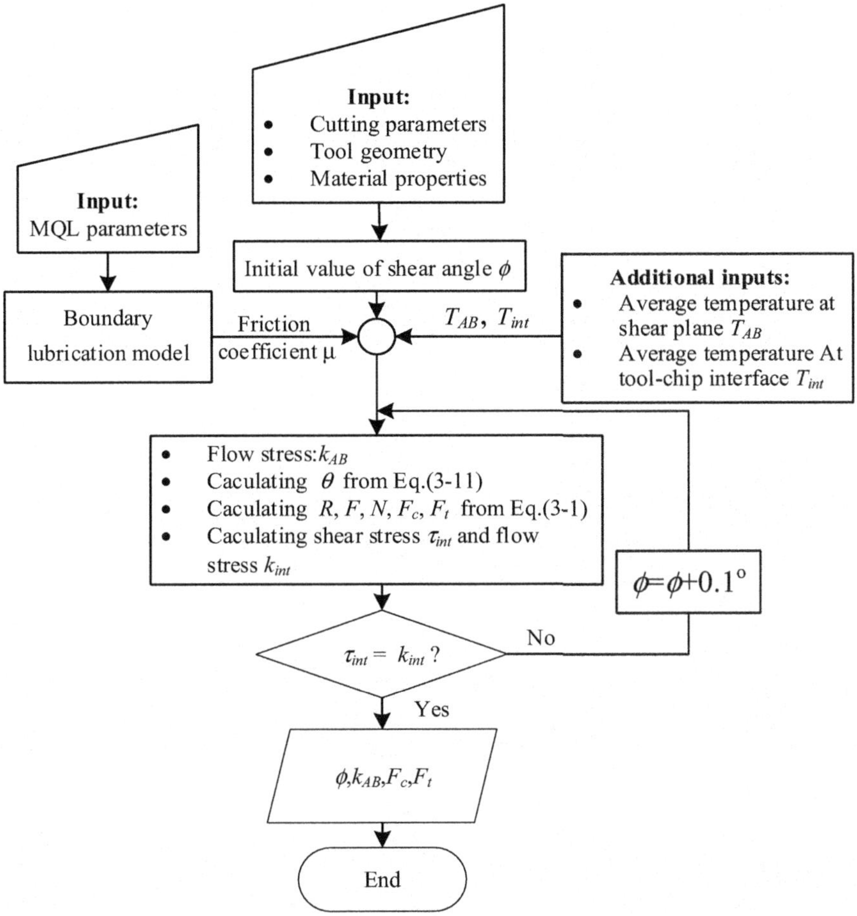

Fig. 3.14 Schematic of the modified Oxley's model II

3.3.2 Iterative Coupling Model of Cutting Force and Cutting Temperature

Based on the modified Oxley's predictive model, cutting force and cutting temperature in MQL machining can be directly predicted according to the cutting parameters, tool geometry, material properties, and MQL parameters. The specific process is shown in Fig. 3.15

Step I, the cutting forces in dry machining is predicted by the cutting force predictive model, which is introduced in Sect. 3.1.

Step II, the obtained cutting force in dry machining is applied into the boundary lubrication model to predict the friction coefficient in MQL machining, which is introduced in Sect. 2.2.

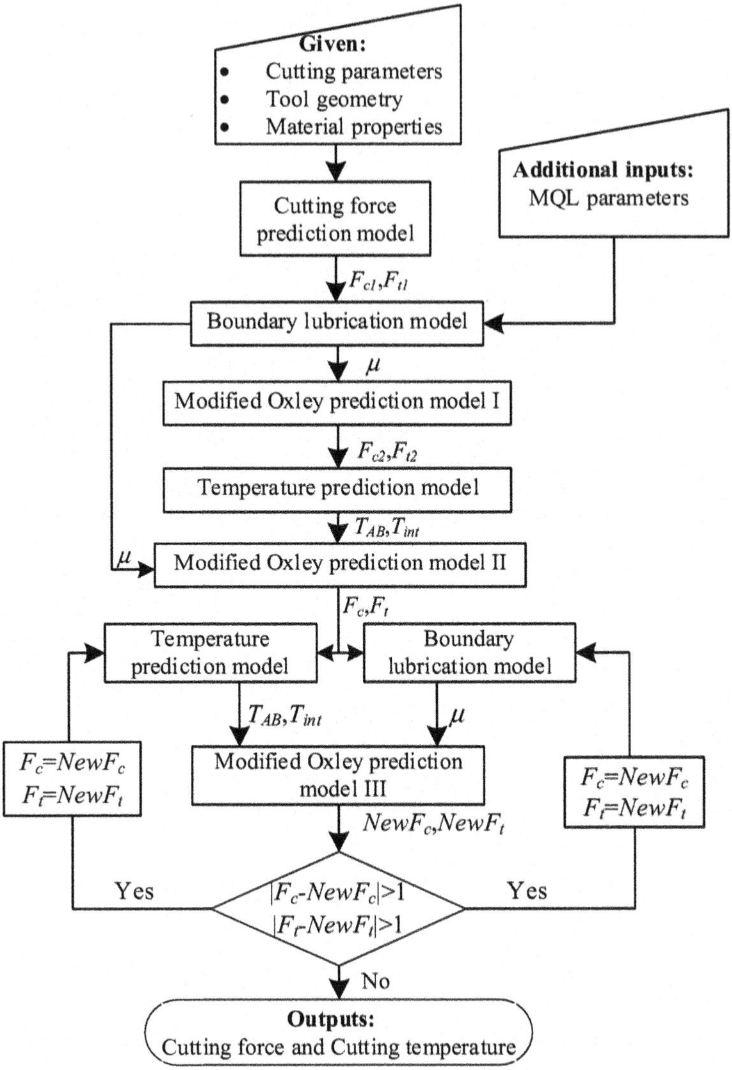

Fig. 3.15 Flow chart of the force and temperature coupled iteration prediction model

Step III, the obtained friction coefficient is applied into the modified Oxley's prediction model I to calculated the cutting force in MQL machining.

Step IV, the obtained cutting force in MQL machining is applied into the cutting temperature model, which is introduced in Sect. 3.2, to predict the average temperature at shear plane T_{AB} and the average temperature at tool-chip interface T_{int}.

Step V, the predicted T_{AB}, T_{int} and the friction coefficient under MQL condition are substituted into the modified Oxley prediction model II. Therefore, the cutting force under lubrication and cooling effects of MQL can be obtained.

Step VI, the predicted cutting forces are taken as the initial value for MQL process, and then it is substituted into the boundary lubrication model and temperature prediction model to predict the coefficient of friction and cutting temperature under MQL condition. The prediction results are substituted into the modified Oxley prediction model II. By iteration, the final cutting force and cutting temperature are achieved until the deviation is within the required range, the flowchart of the above is shown in Fig. 3.15.

3.4 Comparison of Cutting Force Prediction Models

From the discussed above, it is found that the proposed prediction model in this book can be used in dry machining, MQL machining, and flood cooling machining. While Oxley's prediction model is only applicable for dry machining, the prediction model proposed by Li and Liang is only applicable for MQL machining. The details of the comparison of the three prediction models are introduced in the following context.

3.4.1 Comparison of Model Inputs and Application Scope

The input parameters and application scope of the above three prediction models are shown in Table 3.1.

Table 3.1 Comparisons of inputs and cutting conditions for cutting force prediction model

Cutting force model	The model established in this paper	Oxley's prediction model	Li and Liang's model
Model input	1. Cutting parameters 2. Tool geometry 3. Material properties 4. Lubrication conditions	1. Cutting parameters 2. Tool geometry 3. Material properties	1. Cutting parameters 2. Tool geometry 3. Material properties 4. Lubrication conditions 5. Cutting temperature
Scope of application	Dry, MQL, traditional flood cooling	Dry	Dry, MQL

Table 3.2 Cutting parameters and measured cutting forces under dry condition [15]

Test no.	Speed (cm/s)	Feed rate (cm/rev)	Depth of cut (cm)	Measured F_c (N)	Measured F_t (N)
1	76.25	0.00508	0.0508	165.07	99.91
2	76.25	0.00762	0.1016	219.47	141.71
3	76.25	0.01016	0.0762	211.43	134.37
4	152.5	0.00508	0.1016	222.84	219.83
5	152.5	0.00762	0.0762	209.79	185.79
6	152.5	0.01016	0.0508	170.33	134.12
7	228.75	0.00508	0.0762	132.08	100.69
8	228.75	0.00762	0.0508	135.97	103.83
9	228.75	0.01016	0.1016	287.01	168.72

3.4.2 Comparison of Predicted Cutting Force

Taking the cutting of AISI1045 steel with uncoated carbide tool as an example, the tool rake angle is 0° and the clearance angle is 11°. Since Oxley's prediction model is only applicable for dry cutting, the cutting forces predicted by the proposed model in this book are compared with those predicted by Oxley's prediction model. The machining parameters and measured cutting forces under dry cutting conditions are from Ref. [15], as shown in Table 3.2.

The comparison between the prediction results obtained by the two prediction models and the experimental measurement results is shown in Fig. 3.16. For the cutting force in cutting direction, the prediction error of the model proposed in this paper is 0.4–15.9%. In the first group, the maximum prediction error is 57.2%, and the average prediction error is 10.3%. The prediction error obtained by Oxley's prediction model is 1.5–28%, and the average prediction error is 9.6%. For the forces in thrust direction, the prediction error of the model proposed in this paper is 0.9–23.3%, and the average prediction error is 6.7%. The prediction error of Oxley's model is 4.4–29.8%, and the average prediction error is 16%. Based on the comparison of cutting force in cutting direction and thrust direction, it can be seen that the prediction accuracy of cutting force model proposed in this book is higher than that of Oxley's model.

The cutting force measured under MQL condition is compared with the cutting force predicted by the proposed model in this book and predicted by Li and Liang's model [15], and the results are shown in Fig. 3.17. In this book, only cutting forces in thrust direction are compared. From the comparison results in Fig. 3.17, it can be seen that the prediction error of the proposed model in this book is 4.2–18.8%, and the average prediction error is 13.4%. The prediction error of Li and Liang's cutting force prediction model is 4.4–20.9%, and the average prediction error is 10.2%. It can be seen that the prediction accuracy of the cutting force model proposed in this book is equivalent to that of Li and Liang. However, Li and Liang's prediction model needs to measure the cutting temperature, which is difficult to measure in the

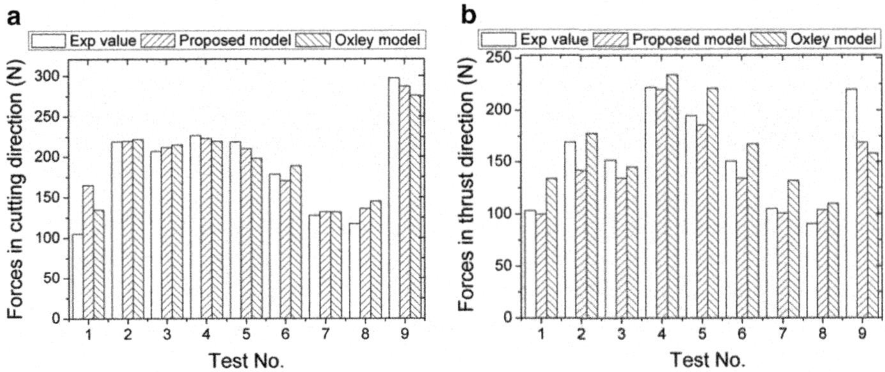

Fig. 3.16 Comparisons of predicted cutting force under dry condition: **a** Forces in cutting direction. **b** Force in thrust direction

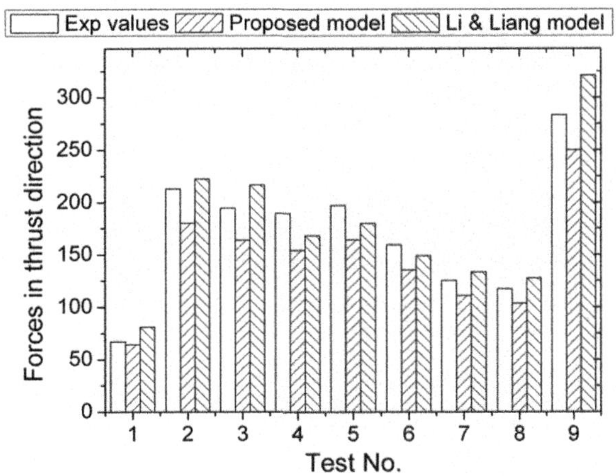

Fig. 3.17 Comparisons of predicted cutting force under MQL condition

actual test process. Based on the above analysis, the cutting force prediction model proposed in this book is superior to other prediction models in both application scope and prediction accuracy.

3.5 Summary

This chapter introduces the coupling prediction model of cutting force and cutting temperature. The coupling is mainly considered from two aspects: firstly, the cutting force and cutting temperature are coupled based on the flow stress model of materials. In the process of cutting, cutting heat is generated by cutting force. According to Johnson–Cook flow stress model, the change of cutting temperature causes the change of material flow stress, which leads to the variation of cutting force. The variation of cutting force also leads to the change of cutting temperature. In this way, the cutting forces and cutting temperatures are coupled and finally achieve a dynamic balance. Secondly, the prediction model for cutting forces and cutting temperatures under MQL conditions are coupled by iteration method. The main contribution is as follows:

(1) In this book, the chip formation force prediction model and plowing force prediction model under MQL condition are established based on Oxley's prediction theory and Waldorf prediction model, which considers the lubrication effect due to MQL. The cutting temperature need to be measured as the prediction model inputs.
(2) The temperature prediction model for the workpiece, tool and chip under MQL conditions is established based on the dry cutting temperature prediction model which is proposed by Komanduri. The model considers the cooling effect due to MQL, and the cutting force needs to be measured as the model inputs. The moving and fixed heat source method proposed by Jager is utilized, and the heat balance equation between chip, tool and workpiece is established.
(3) The coupling prediction model of cutting force and cutting temperature is established based on the modified Oxley's prediction model, which considers the lubrication and cooling effects generated by MQL. The cutting parameters, tool geometry, material properties and MQL parameters are taken as model inputs.
(4) The cutting force prediction model proposed in this book is compared with Oxley's prediction model and Li and Liang's prediction model from the application scope and prediction accuracy. It is indicated that the cutting force prediction model proposed in this book has obvious advantages over the other two models.

References

1. Merchant ME (1945) Mechanics of the metal cutting process II. Plasticity conditions in orthogonal cutting. J Appl Phys 16(6):318–324
2. Oxley PLB (1989) The mechanics of machining: an analytical approach to assessing machinability, 1st edn. Ellis Horwood Ltd., New York
3. Lalwani D, Mehta N, Jain P (2009) Extension of Oxley's predictive machining theory for Johnson and Cook flow stress model. J Mater Process Technol 209(12–13):5305–5312
4. Boothroyd G (1963) Temperatures in orthogonal metal cutting. Proc Inst Mech Eng 177(1):789–810
5. Manjunathaiah J, Endres WJ (2000) A new model and analysis of orthogonal machining with an edge-radi used tool. J Manuf Sci Eng—Trans ASME 122:385–390
6. Arsecularatne J (1997) On tool-chip interface stress distributions, ploughing force and size effect in machining. Int J Mach Tools Manuf 37(7):885–899
7. Waldorf DJ (2006) A simplified model for ploughing forces in turning. J Manuf Process 8(2):76–82
8. Fang N (2003) Slip-line modeling of machining with a rounded-edge tool—Part I: new model and theory. J Mech Phys Solids 51(4):715–742

9. Fang N (2003) Slip-line modeling of machining with a rounded-edge tool—Part II: analysis of the size effect and the shear strain-rate. J Mech Phys Solids 51(4):743–762
10. Bhokse V, Chinchanikar S, Anerao P, Kulkarni A (2015) Experimental investigations on chip formation and plowing cutting forces during hard turning. Mater Today Proc 2(4–5):3268–3276
11. Chung C, Tran MQ, Liu MK (2020) Estimation of process damping coefficient using dynamic cutting force model. Int J Precis Eng Manuf 21:623–632
12. Waldorf DJ, DeVor RE, Kapoor S (1998) A slip-line field for ploughing during orthogonal cutting. J Manuf Sci Eng 120(4):693–698
13. Komanduri R, Hou ZB (2001) Thermal modeling of the metal cutting process—Part II: temperature rise distribution due to frictional heat source at the tool-chip interface. Int J Mech Sci 43(1):57–88
14. Huang Y, Liang SY (2003) Modelling of the cutting temperature distribution under the tool flank wear effect. Proc Inst Mech Eng C J Mech Eng Sci 217(11):1195–1208
15. Li KM, Liang SY (2007) Modeling of cutting forces in near dry machining under tool wear effect. Int J Mach Tools Manuf 47(7–8):1292–1301

Chapter 4
Residual Stress Model in MQL Machining

Based on the coupling prediction model of cutting force and cutting temperature under minimum quantity lubrication (MQL), the prediction model of residual stress in MQL is introduced in details. Firstly, the stress distribution on the workpiece surface is calculated according to the Hertzian contact model based on the predicted cutting force and cutting temperature. Then, the stress distribution is substituted into the modified McDowell algorithm model to calculate the residual stress distribution on the machined surface under MQL condition.

4.1 Prediction of Stress Distribution

The stress distribution at the surface of the machined workpiece is calculated by the Hertzian rolling slip contact theory. As shown in Fig. 4.1, the round object is regarded as the tool and the rectangular object is regarded as the workpieces. In metal cutting, the contact between the tool and the workpiece can be taken as a rolling contact of a cylinder on a semi-infinite plane, as shown in Fig. 4.1.

It is considered that the stress process at any point at the same depth is the same for the rolling-sliding contact model. That is to say, any point with the same depth in the workpiece has the same stress distribution as shown in Fig. 4.1. The stress history only depends on the load on the surface of the workpiece.

The stress distribution at the surface of the machined workpiece is discussed from two aspects: the mechanical load and thermal load.

4.1.1 Stress Due to Mechanical Load

The stress source induced by mechanical load in cutting process mainly consider two aspects: one is the stress induced by shear deformation of the workpiece, mainly

© The Author(s), under exclusive license to Springer Nature Singapore Pte Ltd. 2023

X. Ji, *Minimum Quantity Lubrication Machining*,
https://doi.org/10.1007/978-981-19-7087-0_4

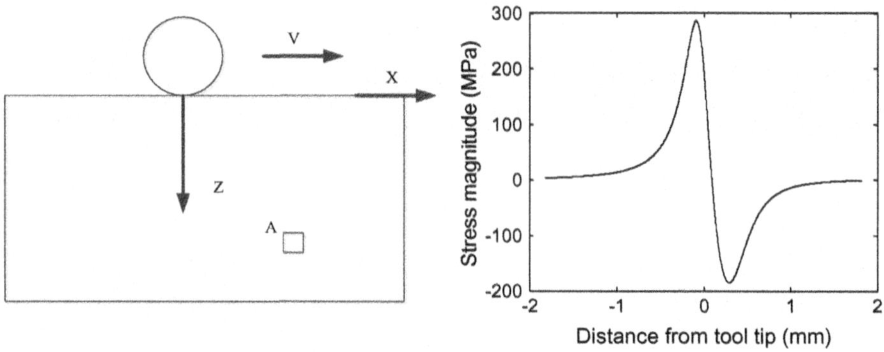

Fig. 4.1 Contact stress load history

including normal stress and shear stress at the shear plane. The other is the stress induced by the friction at tool-workpiece interface, mainly including the normal stress and tangential stress at tool-workpiece interface as shown in Fig. 4.2.

It is assumed that the contact between the tool and the workpiece can be regarded as a Hertzian rolling-sliding contact, so the stress distribution in the workpiece can be derived from the Hertzian stress solution of the rolling-sliding contact model. Figure 4.3 shows that in the narrow band ($-a < x < a$), the normal pressure $p(s)$and tangential stress $q(s)$ are arbitrarily distributed on the elastic half-space, and the stress distribution caused by $p(s)$ and $q(s)$ at any point A in the workpiece can be calculated. As shown in Fig. 4.3, for Point B with a distance s from the original point, the force acting on the area with a width of ds can be considered as pds acting on the normal direction and qds acting on the thrust direction of the surface. The stress component at point A produced by $p(s)$ and $q(s)$ distributions can be derived by integral of the Boussinesq solution within the loaded region ($-a < x < a$):

Fig. 4.2 Stress sources of mechanical load

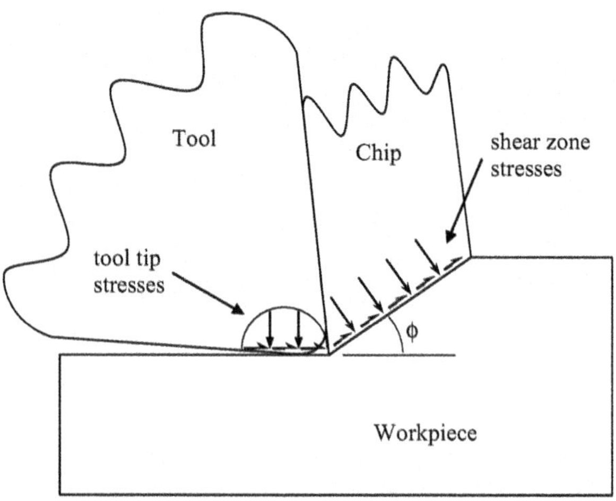

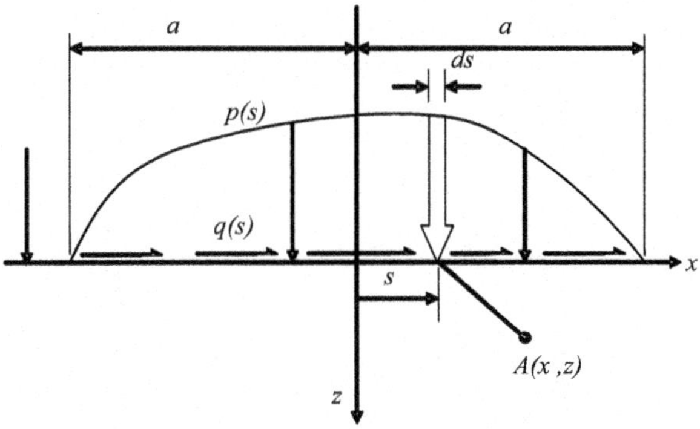

Fig. 4.3 Schematic of boundary stresses [1]

$$\sigma_x = -\frac{2z}{\pi}\int_{-a}^{a}\frac{p(s)(x-s)^2}{\left[(x-s)^2+z^2\right]^2}ds - \frac{2}{\pi}\int_{-a}^{a}\frac{q(s)(x-s)^3}{\left[(x-s)^2+z^2\right]^2}ds$$

$$\sigma_z = -\frac{2z^3}{\pi}\int_{-a}^{a}\frac{p(s)}{\left[(x-s)^2+z^2\right]^2}ds - \frac{2z^2}{\pi}\int_{-a}^{a}\frac{q(s)(x-s)}{\left[(x-s)^2+z^2\right]^2}ds$$

$$\tau_{xz} = -\frac{2z^2}{\pi}\int_{-a}^{a}\frac{p(s)(x-s)}{\left[(x-s)^2+z^2\right]^2}ds - \frac{2z}{\pi}\int_{-a}^{a}\frac{q(s)(x-s)^2}{\left[(x-s)^2+z^2\right]^2}ds \quad (4.1)$$

In order to utilize Eq. (4.1), normal stress distribution $p(s)$ and tangential stress distribution $q(s)$ in Fig. 4.3 need to be calculated firstly. The stress distribution at the tool-workpiece interface and shear plane is introduced in details as following:

(1) Stress due to contact of tool tip and the machined surface

The stress induced by the tool tip and machined surface depends on the ploughing force and the contact area. Assuming that the normal stress at tool-workpiece interface is uniformly distributed, it can be calculated by Eq. (4.2):

$$p_{tool-edge} = \frac{2P_{thrust}}{\pi(wa)} \quad (4.2)$$

where P_{thrust} is the ploughing force in the normal direction, w is the cutting width, a is the contact half width $a = \frac{1}{2}CA$, CA can be calculated by Eq. (3.32).

It is assumed that the thrust stress at tool-workpiece interface is also uniformly distributed and it is proportional to the plowing force in cutting direction P_{cut} and the friction coefficient μ. Therefore, the tangential stress distribution at tool-workpiece interface can be calculated by:

$$\tau_{tool-edge} = \mu\left(\frac{P_{cut}}{w \cdot CA}\right) \quad (4.3)$$

Replacing $p(s)$ and $q(s)$ in Eq. (4.1) with $p_{tool-edge}$ and $\tau_{tool-edge}$, respectively, the stress component induced at tool-workpiece interface can be calculated.

(2) Stress due to shear deformation

Assuming that the normal stress at the shear plane is uniformly distributed, the normal stress distribution at the shear plane can be calculated according to the cutting force and the shear plane by Eq. (4.4):

$$p_s = \frac{F_c \sin \phi + F_t \cos \phi}{L_{AB} w} \tag{4.4}$$

where F_c, F_t are chip formation force, ϕ is shear angle, L_{AB} is the length of shear plane, w is the cutting width.

Since the tangential stress generated by the shear deformation can be regarded as the flow stress of material at shear plane, the shear stress can be calculated by:

$$q_s = k_{AB} \tag{4.5}$$

where k_{AB} is the flow stress of the material, which can be obtained by Eq. (3.10).

Substituting $p(s)$ and $q(s)$ in Eq. (4.1) for normal stress density p_s and tangential stress density q_s respectively, the stress component generated by shear deformation can be calculated.

The total stress in the workpiece due to mechanical load is the sum of the stress due to contact between the cutting edge and the workpiece and the stress due to shear deformation. As the two stress sources are not in the same coordinate system, as shown in Fig. 4.4, the stress component of the shear plane obtained in the coordinate system X'–Z' needs to be converted into the workpiece coordinate system X–Z.

The stress component of the shear plane calculated by Eq. (4.1) is established in the coordinate system X'–Z'. The stress component obtained in the coordinate system X'-Z' needs to be converted into the workpiece coordinate system X–Z through the

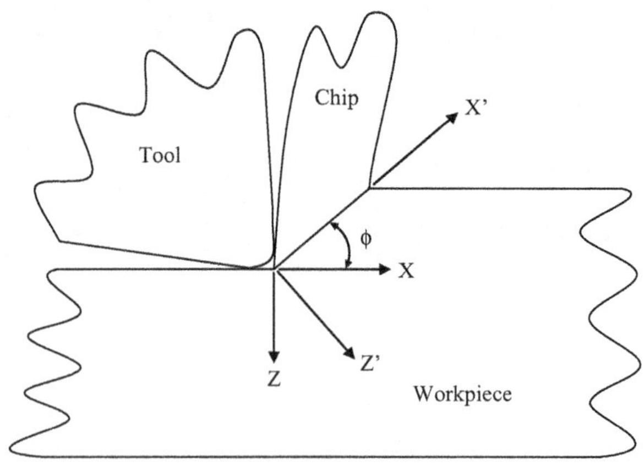

Fig. 4.4 Coordinate system of shear plane with respect to workpiece

transformation matrix by Eq. (4.6).

$$[Q] = \begin{bmatrix} \cos\phi & \sin\phi \\ -\sin\phi & \cos\phi \end{bmatrix} \tag{4.6}$$

Therefore, the stress component caused by shear plastic deformation at any point on the workpiece can be expressed in the workpiece coordinate system X–Z as:

$$\left[\sigma_{shear\_X-Z}\right] = [Q]\begin{bmatrix} \sigma_{X'} & \tau_{X'Z'} \\ \tau_{X'Z'} & \sigma_{Z'} \end{bmatrix}\left[Q^T\right] \tag{4.7}$$

To sum up, for any point in the workpiece, the stress generated by mechanical load is the superposition of the stress component generated by shear deformation and the stress component generated by the contact between tool edge and the machined surface, which can be expressed as:

$$\left[\sigma_{total}^{mech}\right] = \left[\sigma_{shear\_X-Z}\right] + \left[\sigma_{tooledge\_X-Z}\right] \tag{4.8}$$

4.1.2 Stress Due to Thermal Load

The temperature of any point in the workpiece can be predicted by the temperature prediction model. Therefore, the thermal stress caused by temperature at any point in the workpiece mainly includes: (1) thermal stress $X = -(\alpha E/(1-2\upsilon))(\delta T/\delta x)$ caused by volume force and the thermal stress $Z = -(\alpha E/(1-2\upsilon))(\delta T/\delta z)$ along depth; (2) surface tension due to temperature $\alpha ET/(1-2\upsilon)$; (3) Hydrostatic pressure $\alpha ET/(1-2\upsilon)$.

For any point in the workpiece, thermal stress caused by temperature change can be calculated by superposition as following [1, 2]:

$$\begin{aligned}
\sigma_{xx}^{therm}(x,z) &= -\frac{\alpha E}{1-2\upsilon}\int_0^{\infty}\int_{-\infty}^{\infty}\left(G_{xh}\frac{\partial T}{\partial x}(x,z) + G_{xv}\frac{\partial T}{\partial x}(x,z)\right)dxdz \\
&\quad + \frac{2z}{\pi}\int_{-\infty}^{\infty}\frac{p(t)(t-x)^2}{\left((t-x)^2+z^2\right)^2}dt - \frac{\alpha ET(x,z)}{1-2\upsilon} \\
\sigma_{zz}^{therm}(x,z) &= -\frac{\alpha E}{1-2\upsilon}\int_0^{\infty}\int_{-\infty}^{\infty}\left(G_{zh}\frac{\partial T}{\partial x}(x,z) + G_{zv}\frac{\partial T}{\partial x}(x,z)\right)dxdz \\
&\quad + \frac{2z^3}{\pi}\int_{-\infty}^{\infty}\frac{p(t)}{\left((t-x)^2+z^2\right)^2}dt - \frac{\alpha ET(x,z)}{1-2\upsilon} \\
\tau_{xz}^{therm}(x,z) &= -\frac{\alpha E}{1-2\upsilon}\int_0^{\infty}\int_{-\infty}^{\infty}\left(G_{xzh}\frac{\partial T}{\partial x}(x,z) + G_{xzv}\frac{\partial T}{\partial x}(x,z)\right)dxdz
\end{aligned}$$

$$+\frac{2z^2}{\pi}\int_{-\infty}^{\infty}\frac{p(t)(x-t)}{\left((t-x)^2+z^2\right)^2}dt \tag{4.9}$$

where

$$p(t)=\frac{\alpha ET(x,z=0)}{1-2\upsilon} \tag{4.10}$$

where E is the elastic modulus of the material, υ is the Poisson's ratio of the material, α is the thermal diffusion coefficient of the material, $\alpha = \frac{k}{\rho C_p}$. k, ρ and C_p represent the thermal conductivity, density and specific heat capacity of the material respectively. In Eq. (4.9), G_{xh}, G_{xv}, G_{zh}, G_{zv}, G_{xzh} and G_{xzv} respectively represent Green's function of plane strain. For example, $G_{xh}(x, z)$ represents the thermal stress produced by normal stress $\sigma_{xx}(x, z)$ acting in X direction. Similarly, $G_{xv}(x, z)$ represents the thermal stress produced by normal stress $\sigma_{xx}(x, z)$ acting in Z direction.

Therefore, for any point in the workpiece, the thermal stress due to temperature change can be expressed as:

$$\left[\sigma_{total}^{therm}\right]=\begin{bmatrix}\sigma_{xx}^{thermal} & & \\ & \sigma_{zz}^{thermal} & \\ & & \tau_{xz}^{thermal}\end{bmatrix} \tag{4.11}$$

To sum up, the total stress at any point on the workpiece due to mechanical load and thermal load can be calculated by Eq. (4.12):

$$[\sigma_{total}]=\left[\sigma_{total}^{mech}\right]+\left[\sigma_{total}^{thermal}\right] \tag{4.12}$$

4.1.3 Stress Analysis

Taken cutting of AISI4130 alloy steel with uncoated cutting tool as example, the rake angle is 5°, and the clearance angle is 11°, the radius of the tool edge is 30 μm. The cutting speed is 1.049 m/s, the cutting depth is 0.0508 mm, the cutting width is 4.775 mm, the flow rate of MQL is 16 ml/h. When the gas pressure is 40 psi, the predicted temperature distribution in the workpiece is shown in Fig. 4.5, and its stress distribution is shown in Fig. 4.6.

From Fig. 4.5, it is found that when the cutting tool approaches the workpiece from far away, the temperature in the workpiece increases rapidly and reaches the maximum value. When the tool is far away from the workpiece, the temperature in the workpiece decreases slowly. It is also found that the peak temperature in the workpiece is not beneath the tool tip, but it is close to the tool tip, which is caused by the heat conduction of the workpiece material.

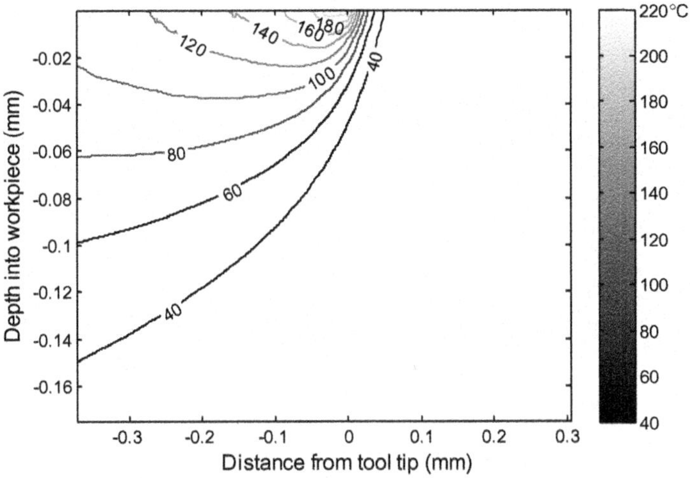

Fig. 4.5 The predicted temperature distributions in workpiece

From Fig. 4.6, it is found that the stress component caused by shear deformation is dominated, and the stress caused by shear deformation is significantly greater than that caused by the contact between the tool tip and the machined surface, especially for the normal stress along X direction and the shear stress in the X–Z plane. From Fig. 4.6b, for the normal stress along Z direction, the stress caused by shear deformation is equivalent to that caused by the tool tip and machined surface. This is because in this cutting condition, the tool edge radius is 30 μm, which is close to the cutting depth 0.0508 mm, so the stress caused by the tool edge radius cannot be ignored.

4.2 Residual Stress Prediction Model

The obtained surface stress of the workpiece calculated in Sect. 4.1 is applied into the incremental thermal-elastic–plastic model to predict the residual stress. The load applied to the workpiece surface due to the tool can be regarded as a moving load, and it is a cyclic process. Many scholars used incremental plastic model to predict residual stress generated by rolling contact surfaces [3–5]. The Merwin model assumes that the workpiece material is an ideal elastic–plastic material and an elastic strain field is utilized to predict the residual stress on the rolling contact surface [3]. Considering the properties of the materials in machining, the incremental plasticity theory based on the kinematic hardening model proposed by McDowell is used to predict the residual stress at the machined surface.

Fig. 4.6 Stress distribution in workpiece

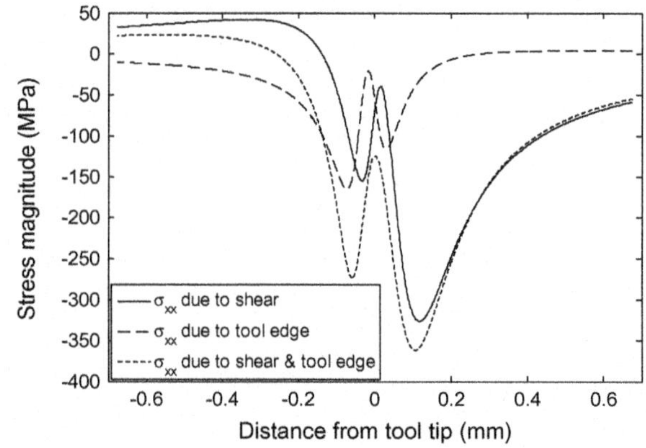

(a) Normal stress in the X direction

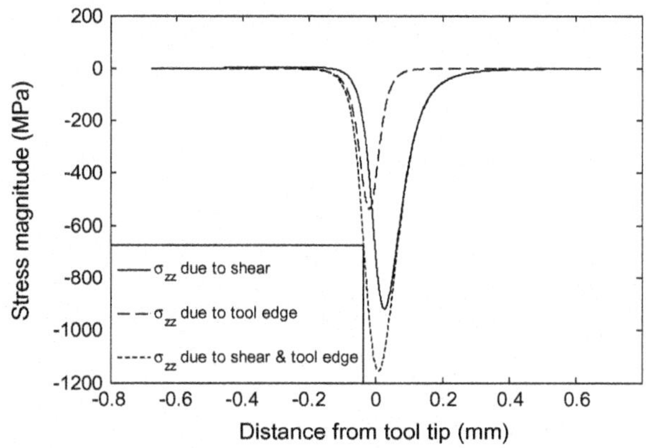

(b) Normal stress in the Z direction

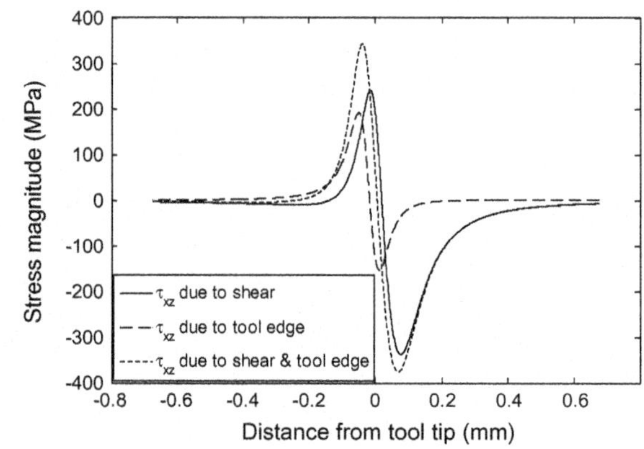

(c) Shear stress in the X-Z plane

4.2.1 Model Criteria

(1) Material strengthening criteria

Because the metal cutting process is accompanied by high temperature, high pressure, large strain and high strain rate, the isotropic strengthening model which does not consider the Bauschinger effect is not suitable for predicting the subsequent yield stress in metal cutting process. The kinematic strengthening model is utilized to predict the variation of yield stress under plastic strain, which is more consistent with the properties of material in actual machining process.

(2) Material yield criterion

Tresca yield criterion needs to know the magnitude and direction of the principal stress at each point in the workpiece, while only the stress component at any point in the workpiece can be calculated according to the introduction in Sect. 4.1. Therefore, Von-Mises yield criterion is more suitable to judge whether material is yielded or not during machining process.

The process of stress loading and relax during metal cutting is described in details to predict the final residual stress distribution in the workpiece.

4.2.2 Load History

Firstly, the stress cyclic load history is introduced. For any point in the workpiece, the stress loading process can be divided into N steps. The effective strain and effective strain rate can be calculated by Eqs. (4.13) and (4.14).

$$\varepsilon_{eff}^{p} = \frac{\sqrt{2}}{3}\sqrt{\left(\varepsilon_{xx}^{p} - \varepsilon_{yy}^{p}\right)^{2} + \left(\varepsilon_{yy}^{p} - \varepsilon_{zz}^{p}\right)^{2} + \left(\varepsilon_{zz}^{p} - \varepsilon_{xx}^{p}\right)^{2} + 6\left(\varepsilon_{xz}^{p}\right)^{2}} \tag{4.13}$$

$$\dot{\varepsilon}_{eff}^{p} = \sqrt{\frac{2}{3}}\sqrt{\left(\dot{\varepsilon}_{xx}^{p}\right)^{2} + \left(\dot{\varepsilon}_{yy}^{p}\right)^{2} + \left(\dot{\varepsilon}_{zz}^{p}\right)^{2} + 2\left(\dot{\varepsilon}_{xz}^{p}\right)^{2}} \tag{4.14}$$

According to the calculated effective strain and effective strain rate, Johnson–Cook flow stress model is used to calculate the yield stress of material under transient condition, and Von-Mises yield criterion is adopted to judge whether the workpiece material is yielded or not. The yield criterion can be expressed as follows:

$$F = \frac{3}{2}\left(S_{ij} - \alpha_{ij}\right)\left(S_{ij} - \alpha_{ij}\right) - R^{2} = 0 \tag{4.15}$$

where deviatoric stress can be obtained by $S_{ij} = \sigma_{ij} - (\sigma_{kk}/3)\delta_{ij}$. R is the shear yield strength of the workpiece, and δ_{ij} is the Kronecker symbol, which is expressed as:

$$\delta_{ij} = \begin{cases} 1 & i = j \\ 0 & i \neq j \end{cases} \tag{4.16}$$

In Eq. (4.15), α_{ij} is the back stress, according to the kinematic strengthening criterion, $\alpha_{ij} = \langle S_{kl} n_{kl} \rangle n_{ij}$. $\langle\rangle$ stands for the MacCauley symbol, and it is defined as $\langle x \rangle = 0.5(x + |x|)$. During the first step of loading, it is assumed that the material does not yield, so $\alpha_{ij} = 0$. n_{ij} is the unit normal vector in the direction of plastic strain rate, which can be expressed as:

$$n_{ij} = \frac{S_{ij} - \alpha_{ij}}{\sqrt{2}k} \tag{4.17}$$

Von-Mises yield criterion is adopted to judge whether the material is yielded or not. If it is not, elastic loading can be carried out and the calculation needs to carried out according to the stress component obtained in Sect. 4.1. If the material is yielded, plastic loading is performed. The plastic flow criterion on the yield surface, namely surface $F = 0$, can be expressed by Eq. (4.18):

$$\dot{\varepsilon}_{ij}^{p} = \frac{1}{h} \langle \dot{S}_{kl} n_{kl} \rangle n_{ij} \tag{4.18}$$

where h represents the plastic modulus, which determines the hardening rate of the material.

Due to the uncertainty of the change of the load in the cutting process, there may be a large fluctuation range. In order to adapt to the change of the load in the cutting process in a large range, McDowell algorithm introduces a mixing function Ψ:

$$\Psi = 1 - \exp\left(-\zeta \frac{3}{2} \frac{h}{G}\right) \tag{4.19}$$

where ζ is the algorithm constant. G is the elastic shear modulus, and it can be calculated by Eq. (4.20):

$$G = \frac{E}{2(1 + \upsilon)} \tag{4.20}$$

where E is elastic modulus, and υ is Poisson's ratio. In this hybrid algorithm, for the plastic flow of workpiece material, neither the strain rate in the cutting direction is assumed to be $\dot{\varepsilon}_{xx} = 0$ [6] nor $\dot{\sigma}_{xx} = \dot{\sigma}_{xx}^{*}$ [7] in the cutting direction. When the stress is plastic loading, the strain rate in the cutting direction satisfies Eq. (4.21). Similarly, since plane strain occurs in Y direction, the strain rate in the Y direction satisfies Eq. (4.22):

$$\begin{aligned} \dot{\varepsilon}_{xx} &= \frac{1}{E}[\dot{\sigma}_{xx} - \nu(\dot{\sigma}_{yy} + \dot{\sigma}_{zz}^{*})] + \alpha\Delta T + \frac{1}{h}(\dot{\sigma}_{xx} n_{xx} + \dot{\sigma}_{yy} n_{yy} + \dot{\sigma}_{zz}^{*} n_{zz} + 2\dot{\tau}_{xz}^{*} n_{xz}) n_{xx} \\ &= \Psi\left(\frac{1}{E}[\dot{\sigma}_{xx}^{*} - \nu(\dot{\sigma}_{yy} + \dot{\sigma}_{zz}^{*})] + \alpha\Delta T + \frac{1}{h}(\dot{\sigma}_{xx}^{*} n_{xx} + \dot{\sigma}_{yy} n_{yy} + \dot{\sigma}_{zz}^{*} n_{zz} + 2\dot{\tau}_{xz}^{*} n_{xz}) n_{xx}\right) \end{aligned} \tag{4.21}$$

$$\dot{\varepsilon}_{yy} = \frac{1}{E}\left[\dot{\sigma}_{yy} - \nu\left(\dot{\sigma}_{xx} + \dot{\sigma}^*_{zz}\right)\right] + \alpha\Delta T + \frac{1}{h}\left(\dot{\sigma}_{xx}n_{xx} + \dot{\sigma}_{yy}n_{yy} + \dot{\sigma}^*_{zz}n_{zz} + 2\dot{\tau}^*_{xz}n_{xz}\right)n_{yy} = 0 \tag{4.22}$$

where $\dot{\varepsilon}_{xx}$ and $\dot{\varepsilon}_{yy}$ represent the strain rate along the cutting direction and the direction which is perpendicular to the cutting direction. $\dot{\sigma}^*_{xx}$, $\dot{\sigma}^*_{yy}$, $\dot{\sigma}^*_{zz}$ and $\dot{\tau}^*_{xz}$ represent the Hertzian stress solution under elastic loading, which can be calculated by Eq. (4.12). $\dot{\sigma}_{xx}$, $\dot{\sigma}_{yy}$ and $\dot{\sigma}_{zz}$ represent the change rate of stress along the cutting direction, direction which is perpendicular to the cutting direction, and the depth direction respectively. According to Eqs. (4.21) and (4.22), the change rate of the stress $\dot{\sigma}_{xx}$ and $\dot{\sigma}_{yy}$ can be calculated. The stress in the workpiece during the whole load history can be calculated by integrating the whole loading path.

4.2.3 Relax Process

After loading, according to the model proposed by Merwin and Johnson, stress and strain should satisfy the boundary conditions [3]:

$$\begin{array}{cccc} (\varepsilon_x)_r = 0 & (\sigma_x)_r = f_1(z) & \left(\varepsilon_y\right)_r = 0 & \left(\sigma_y\right)_r = f_2(z) \\ (\varepsilon_z)_r = f_3(z) & (\sigma_z)_r = 0 & (\gamma_{xz})_r = f_4(z) & (\tau_{xz})_r = 0 \end{array} \tag{4.23}$$

In order to satisfy the above boundary conditions, a stress relax process is carried out at the end of each cyclic load history. For any non-zero components, such as σ^R_{zz}, τ^R_{xz}, ε^R_{xx} and T^R need to be gradually released to 0. Assuming that the release process is divided into M steps, the released components of each step can be expressed as:

$$\Delta\sigma_{zz} = -\frac{\sigma^R_{zz}}{M}, \qquad \Delta\tau_{xz} = -\frac{\tau^R_{xz}}{M}, \qquad \Delta\varepsilon_{xx} = -\frac{\varepsilon^R_{xx}}{M}, \qquad \Delta T = -\frac{T^R}{M} \tag{4.24}$$

Similarly, the Von-Mises yield criterion is also used to determine whether the workpiece material is yielded or not during the relax process. The stress relax process of the material can be divided into two cases: complete elastic relax and plastic relax. For perfect elastic release, it needs to satisfy either $F < 0$ or F=0 and $dS_{ij}n_{ij} \geq 0$. For the elastic relax process, the relax stress increment can be calculated by Eq. (4.25) according to Hooke's Law:

$$\begin{cases} \Delta\sigma_{xx} = \dfrac{E\Delta\varepsilon_{xx} + (1+\nu)(\Delta\sigma_{zz}\nu - E\alpha\Delta T)}{\left(1-\nu^2\right)} \\ \Delta\sigma_{yy} = \dfrac{\nu E\Delta\varepsilon_{xx} + (1+\nu)(\Delta\sigma_{zz}\nu - E\alpha\Delta T)}{\left(1-\nu^2\right)} \end{cases} \tag{4.25}$$

For the plastic relax process, according to Eqs. (4.21) and (4.22), the increment in the stress release process can be obtained:

$$\begin{cases} \Delta\sigma_{yy} = \dfrac{\left(-\frac{\upsilon}{E} + \frac{1}{h}n_{xx}n_{yy}\right)(C - \alpha\Delta T) - \left(\frac{1}{E} + \frac{1}{h}n_{xx}n_{xx}\right)(D - \alpha\Delta T)}{\left[\left(-\frac{\upsilon}{E} + \frac{1}{h}n_{xx}n_{yy}\right)^2 - \left(\frac{1}{E} + \frac{1}{h}n_{xx}n_{xx}\right)\left(\frac{1}{E} + \frac{1}{h}n_{yy}n_{yy}\right)\right]} \\ \Delta\sigma_{xx} = \dfrac{D - \left(\frac{1}{E} + \frac{1}{h}n_{yy}n_{yy}\right)\Delta\sigma_{yy} - \alpha\Delta T}{-\frac{\upsilon}{E} + \frac{1}{h}n_{xx}n_{yy}} \end{cases} \quad (4.26)$$

where

$$\begin{cases} C = \Delta\varepsilon_{xx} + \left(\dfrac{\upsilon}{E} - \dfrac{1}{h}n_{xx}n_{zz}\right)\Delta\sigma^*_{zz} - \dfrac{2}{h}\Delta\tau^*_{xz}n_{xz}n_{xx} \\ D = \left(\dfrac{\upsilon}{E} - \dfrac{1}{h}n_{zz}n_{yy}\right)\Delta\sigma^*_{zz} - \dfrac{2}{h}\Delta\tau^*_{xz}n_{xz}n_{yy} \end{cases} \quad (4.27)$$

The stress and strain obtained by the above stress relax process can be considered as the additional residual stress and strain in the cyclic load history, and the stress and strain obtained by the above flow law can be added as the residual stress and strain of this loading step. The residual stress obtained by the previous loading step can be considered as the initial stress of the next loading step, and it repeats until the stress and strain achieve a stable state, and the final residual stress in the machined surface can be achieved.

4.3 Comparison of Residual Stress Prediction Models

The comparison of two residual stress prediction models is as following: (1) The residual stress prediction models based on the thermal–mechanical effects which is proposed in this book. (2) The residual stress prediction models based on McDowell algorithm. The assumptions and the results of the prediction models are discussed in details as following.

4.3.1 Comparison of Assumption of Two Prediction Models

The comparison of assumptions between the two residual stress prediction models is depicted in Table 4.1.

Table 4.1 Comparisons of assumptions for residual stress prediction model

Prediction models	The residual stress prediction models proposed in this book	The residual stress prediction models based on the McDowell algorithm
Assumptions	1. The cutting process is a steady state, which is approximated as plane strain 2. The contact between the workpiece and the tool is hertz rolling and sliding contact 3. When plastic deformation occurs in cutting process, the kinematic hardening criterion is followed 4. Mechanical load and thermal load is considered, and the effect of phase change of materials on residual stress is not considered	1. The cutting process is a steady state, which is approximated as plane strain 2. The contact between the workpiece and the tool is hertz rolling and sliding contact 3. When plastic deformation occurs in cutting process, the kinematic hardening criterion is followed 4. Mechanical load is considered, and the effect of cutting temperature and phase change of materials on residual stress is not considered

4.3.2 Comparison of Results of the Prediction Models

The processing of alloy steel AISI4130 with uncoated tool is taken as an example. The tool rake angle is 5°. The clearance angle of the tool is 11°. The radius of the tool edge is 30 μm. The cutting speed is 2.098 m/s, the cutting depth is 0.0508 mm, and the cutting width is 4 mm. The flow rate of MQL is 16 ml/h, the air pressure is 40psi.

According to the two residual stress prediction models, the predicted results of residual stress are shown in Figs. 4.7 and 4.8. The residual stress predicted by the proposed model in this book is shown in Fig. 4.7, which considers the thermal effect on the residual stress. It is found that the residual stress at the surface of the workpiece is 457.5 MPa in the cutting direction and −61.5 MPa in the radial direction. The residual stress predicted by McDowell's algorithm is shown in Fig. 4.8. It is found that the residual stress at the surface of the workpiece is −186.5 MPa in the cutting direction and −382.0 MPa in the radial direction. Both the cutting direction and the radial direction are all compressive residual stress. From the comparison between Figs. 4.7 and 4.8, it is found that the influence of mechanical load on compressive residual stress is mainly manifested as compressive residual stress, while the influence of thermal load on residual stress is mainly manifested as tension residual stress.

4.4 Summary

The prediction of residual stress on workpiece surface generated by mechanical load and thermal load in the cutting process of MQL is discussed in details. The main contribution is as follows:

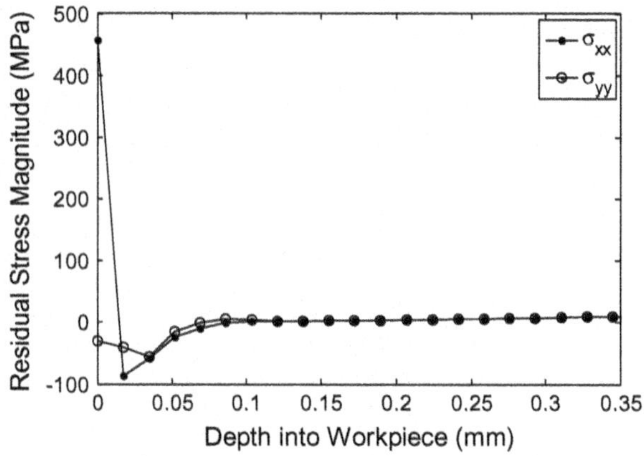

Fig. 4.7 Predicted results by proposed prediction model

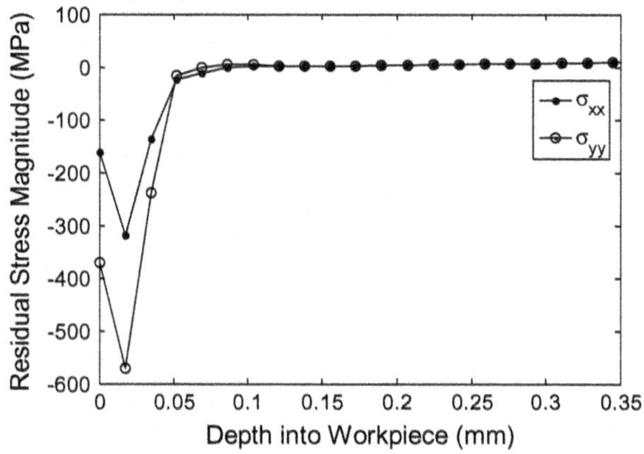

Fig. 4.8 Predicted results by McDowell model

(1) According to the cutting force, cutting temperature and the geometry of the cutting area predicted in Chap. 3 in MQL machining, the stress distribution in the workpiece due to mechanical load and thermal load can be calculated by the Hertz rolling and sliding contact model, and the stress distribution in the workpiece is analyzed.
(2) The predicted stress component of the workpiece is put into the increment thermal-elastic–plastic model and the McDowell mixed algorithm is adopted. The kinematic hardening of workpiece material under the load changing widely in the MQL cutting process is applied, final residual stress distribution on workpiece surface is predicted through the stress load process and stress relax process.
(3) From the comparison of two residual stress prediction models, it can be concluded that the residual stress caused by thermal load is mainly manifested as tension residual stress, while the residual stress caused by mechanical load is mainly manifested as compressive residual stress.

References

1. Johnson KL (1985) Contact mechanics. Cambridge University Press, Cambridge, New York
2. Saif M, Hui C, Zehnder A (1993) Interface shear stresses induced by non-uniform heating of a film on a substrate. Thin Solid Films 224(2):159–167
3. Merwin J, Johnson KL (1963) An analysis of plastic deformation in rolling contact. Proc Inst Mech Eng 177(1):676–690
4. McDowell D (1997) An approximate algorithm for elastic-plastic two-dimensional rolling/sliding contact. Wear 211(2):237–246
5. Ji X, Li B, Liang SY (2018) Analysis of thermal and mechanical effects on residual stress in Minimum Quantity Lubrication (MQL) machining. J Mech 34(1):41–46
6. McDowell DL, Moyar GJ (1986) A more realistic model of nonlinear material response: application to elastic-plastic rolling contact. In: Proceedings of the 2nd international symposium on contact mechanics and wear of rail/wheel systems, Kingston, RI, July 8–11
7. Jiang Y, Sehitoglu H (1994) Analytical approach to elastic-plastic stress analysis of rolling contact. J Tribol, Trans ASME 116(3):577–587

Chapter 5
Experimental Validation by Orthogonal Cutting of AISI 4130 Alloy

In order to verify the prediction model of MQL introduced in the previous chapters, the orthogonal turning test of AISI4130 alloy steel is conducted, the cutting force, cutting temperature and residual stress on the surface of the workpiece during the cutting process is measured. Comparisons are made between the experimental results and prediction results.

5.1 Experimental Method

5.1.1 Design of the Workpiece

In order to keep consistent with the prediction model of orthogonal cutting introduced in the previous chapters, thin-walled cylindrical part is selected for end face turning. The cutting material is AISI4130 alloy steel, with an outer diameter of 63.5 mm, a tube wall thickness of 4.775 mm, and a length of 60 mm. The geometric parameters are shown in Fig. 5.1. The yield strength of the material is 482.65 MPa, and the hardness is HRC19. The chemical composition and content of each element are shown in Table 5.1 [1]. Before machining, the upper and lower surfaces of the workpiece are ground and annealed to relieve stress.

5.1.2 Design of the Setup

The AISI4130 alloy steel tube was turned on the CNC lathe (Hardinge T42SP) with different cutting speeds and feed rates, as shown in Fig. 5.2. Hardinge T42SP is a precision CNC lathe with high power 11 KW, high precision with positioning accuracy and rotation accuracy of 1 μm, high speed with 6000 r/min, high stiffness and good dynamic performance. The lubrication conditions of machining are dry

© The Author(s), under exclusive license to Springer Nature Singapore Pte Ltd. 2023

X. Ji, *Minimum Quantity Lubrication Machining*,
https://doi.org/10.1007/978-981-19-7087-0_5

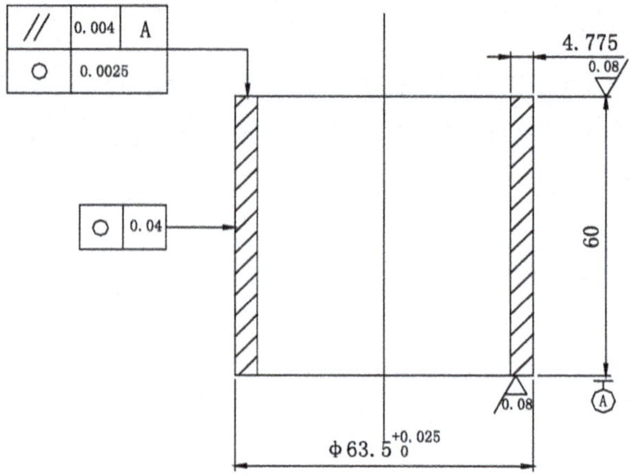

Fig. 5.1 Geometry of the workpiece

Table 5.1 Chemical composition of AISI4130 alloy steel [1]

Element	C	Si	Mn	S	P	Cr	Ni	Cu	Mo
Content (%)	0.26–0.34	0.2–0.4	0.4–0.7	≤0.035	≤0.035	0.8–1.1	≤0.03	≤0.03	0.15–0.25

cutting, MQL cutting and traditional flood cooling machining. The range of cutting speed and feed selection is determined by the tool supplier's recommendation. The handle used in the test was Sandvik conventional handle LC123K08, and the rake angle of the tool is 5°. The tool type is N123K2-0600-0004-Cr 1125, with uncoated carbide insert, the clearance angle of the tool is 11°, the tool edge length is 6 mm, the tool edge radius is 30 μm.

COOLUBRICATOR produced by UNIST is used as the MQL supply system in this experiment, as shown in Fig. 5.3. This system is an independent volumetric continuous spray system. The output flow can be controlled by adjusting the metering

Fig. 5.2 Experimental setup

Fig. 5.3 MQL supply system

pump on the MQL device. The metering pump can adjust the density of air atomization out from the nozzle and the distance and angle of injection. The angle of injection is 15°–20° and the spray area can be controlled by adjusting the position of the MQL nozzle. The pressure supply is continuous during the MQL cutting process, and the air pressure during the test is 40 psi. Coolube2210 vegetable lubricant produced by UNIST is used as cutting fluid. The cutting fluid is a biodegradable synthetic based on natural esters, which is harmless to human body and the environment. In the actual process of machining, the influence of chip on MQL was avoided by adjusting the position of MQL nozzle by experimental observation. In this experiment, the air pressure remains unchanged, and the flow rate of MQL and the air-oil mixture ratio are changed to study the influence of MQL parameters on cutting performance. The flow rate of the MQL supply system can be adjusted from 0 to 250 mL/h, which is realized by the metering pump on the MQL supply system. There are three levels of the air-oil mixture ratio, they are high, medium and low, respectively.

5.1.3 Design of the Measurement

(1) Measurement of cutting force

During the experiment, the cutting force was measured by piezoelectric dynamometer (Kistler 9257B). The installation position of the dynamometer is shown in Fig. 5.4. First, the dynamometer is installed on the lathe through the fixture, and then the tool rod is installed on the dynamometer through the special fixture. During the cutting process, the cutting force signal is collected by the data acquisition software through the charge amplifier. The whole test system consists of piezoelectric three-dimensional dynamometer, multi-channel charge amplifier, data acquisition card and computer.

As shown in Fig. 5.5, the cutting force collected by data acquisition software Dyno Ware includes three directions: main cutting force F_c in the cutting direction, thrust force F_t, which is perpendicular to the direction of the machined surface, and radial force F_r. The ratio of the wall thickness to the outer diameter of the workpiece

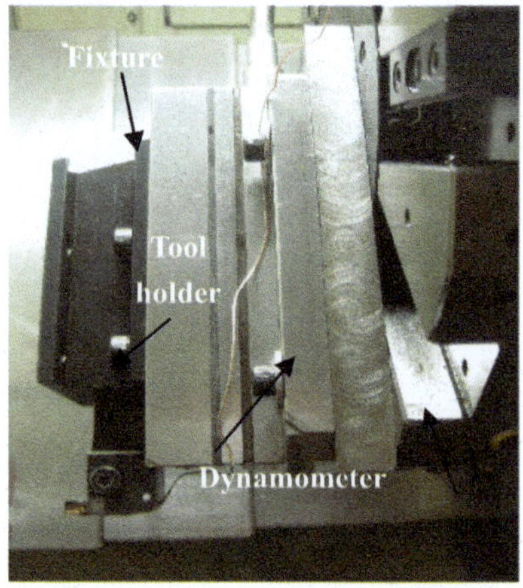

Fig. 5.4 Setup of dynamometer

Fig. 5.5 Illustration of cutting force measurement

material is 1:15, the cutting process can be approximately considered as orthogonal cutting, and the radial force F_r can be ignored in theory.

(2) Cutting temperature measurement

In this experiment, the cutting temperature was measured by thermocouple and thermal imager. The thermocouple was fixed at 4.3 mm from the flank tool surface by thermal synthetic adhesive, as shown in Fig. 5.6. The cutting temperatures of dry cutting, MQL and high-flow lubrication machining were measured respectively. The thermal imager uses FLIR SC6000 thermal imager. Put the thermal imager in front of the lathe with a tripod, as shown in Fig. 5.7. In the actual machining experiment, the safety operation door of the lathe must be closed for the safety under traditional flood cooling machining because the infrared ray cannot penetrate the glass. Therefore, the infrared thermal imager only measured the cutting temperature under dry and MQL conditions.

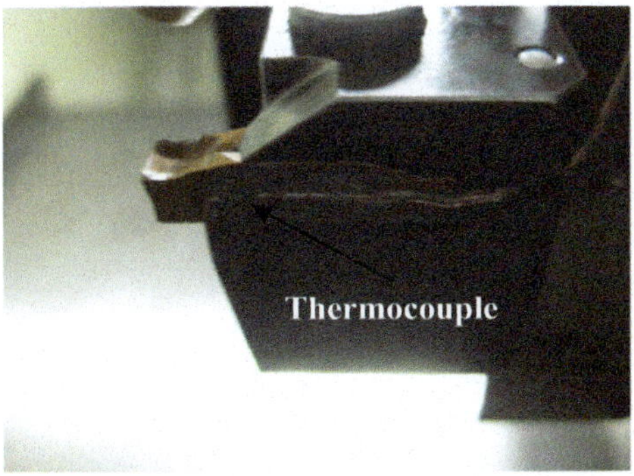

Fig. 5.6 Position of thermal couple

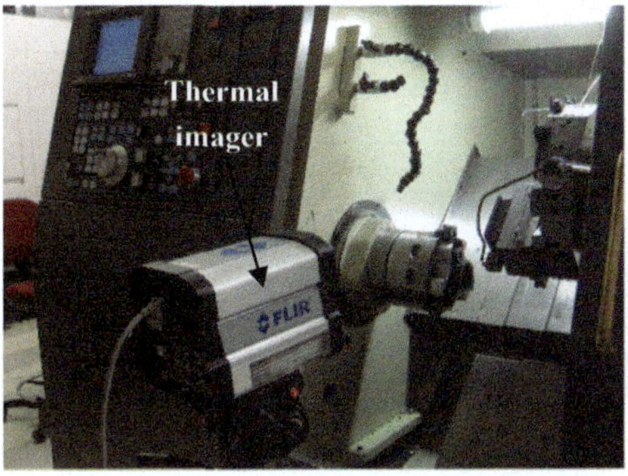

Fig. 5.7 Location of infrared thermal camera

(3) Residual stress measurement

The residual stress in the workpiece was measured by Proto LXRD X-ray diffraction stress dynamometer, as shown in Fig. 5.8. The LXRD dynamometer system can provide a higher level of flexibility to the user. The target material of X-ray is CrKa, the working voltage of X-ray tube is 30.0 kV, the working current of X-ray tube is 25.0 mA, and rectangular spot of 1 mm $\times$ 3 mm is adopted.

As shown in Fig. 5.9, the residual stresses at the machined surface along the cutting direction and the radial direction are measured respectively. The residual stress distribution along the depth direction of the workpiece is measured by electrolytic etching layer by layer at the surface of the workpiece with saturated salt water. Since the workpiece surface is obtained by end face turning, the stress on the workpiece surface can be considered to be a constant at the same depth. Six points on the workpiece surface are selected to be corroded for 1s, 3s, 5s, 10s, 15s and 20s with the current of 3 A, so as to obtain different corrosion depths. Three-dimensional surface optical profilometer (KS1100) is used to measure the corrosion depths of each point, as shown in Fig. 5.10.

Fig. 5.8 X-ray diffractometer

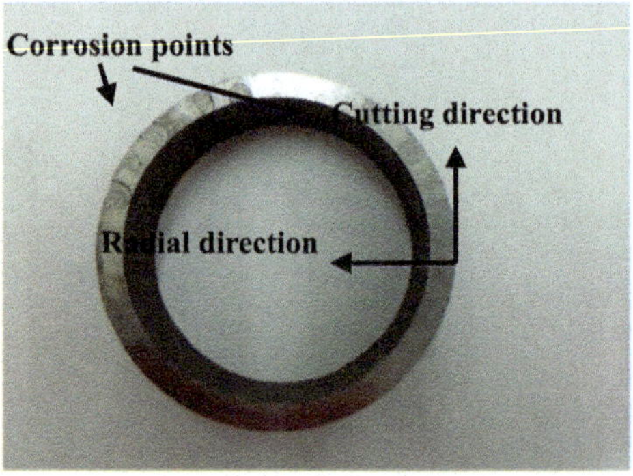

Fig. 5.9 The schematic of residual stress measurement

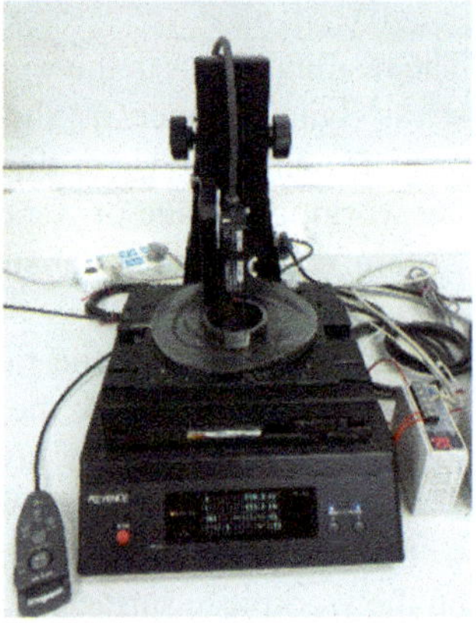

Fig. 5.10 Schematic of corrosion depth measurement

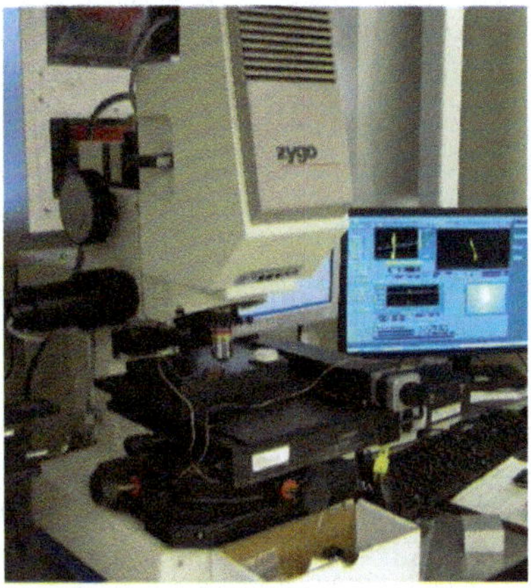

Fig. 5.11 Schematic of tool measurement

(4) Tool surface topography measurement

It is necessary to measure the tool surface topography before determining the friction coefficient in the process of MQL, so as to obtain some parameters of the boundary lubrication model. Before the cutting experiment, Zygo NewView200 was used to measure the tool surface topography (as shown in Fig. 5.11), and the distribution density of microscopic rough points on the tool surface, the height of the microscopic rough point is about 5 μm.

5.2 Analysis of Experimental Results and Model Verification

5.2.1 *Determination of the Parameter in the Prediction Model*

(1) Determination of parameters in boundary lubrication model

Some parameters in the boundary lubrication model need to be determined before calculating the friction coefficient under MQL condition by Eq. (2.17). Firstly, the shear strength of the contact area of the absorbing lubrication film is calculated according to Eq. (5.1) [2]:

$$s_b = \rho \dot{\gamma} \nu \tag{5.1}$$

where, ρ is the density of the lubrication film, ν is the kinematic viscosity of the lubrication film, and $\dot{\gamma}$ is the shear strain rate of the lubrication film. It is assumed that the shear strain rate of the lubrication film is approximately equal to the shear strain rate of the chip, and the shear strain rate of the chip is calculated according to Eq. (3.7).

According to the analysis in Sect. 2.1, the parameters C_1 in Eq. (2.17) is friction coefficient under dry cutting condition, so it can be obtained according to Oxley prediction model. The coefficient C_2 is used to calculate the average contact pressure in the region of absorbing lubrication film, which is assumed to be 0.5 in this analysis according to Kato's boundary lubrication model [3]. The coefficient C_3 is calculated by Eq. (2.8).

The total number of microscopic rough points on the tool and chip interface is predicted by Eq. (5.2) [3]:

$$n_0 = A_0 z^2 \tag{5.2}$$

where, z represents the linear distribution density of microscopic rough points at tool-chip interface, which can be measured by optical interferometer (Zygo NewView 200) as 100 mm^{-1}. A_0 represents the contact area between the tool and the chip, which can be calculated according to the contact length between the tool and the chip and the cutting width:

$$A_0 = hw \tag{5.3}$$

where, h represents the contact length of the tool and chip, and w represents the cutting width.

The height of the microscopic rough point can be measured by an optical interferometer, which is 5 μm. Table 5.2 shows the parameters which is used to predict the friction coefficient under MQL condition:

(2) Determination of material parameters

Table 5.3 shows the thermal conductivity, specific heat capacity, and density of the AISI4130 alloy steel.

In Sect. 3.1, the Johnson–Cook flow stress model is used to predict the flow stress during cutting. The Johnson–Cook material constants of AISI4130 alloy steel are shown in Table 5.4:

Table 5.5 shows the parameters of tool material.

Table 5.2 Parameters in boundary lubrication calculation

Parameter	t_b (μm)	C_2	H_{max} (μm)	R (μm)	D	z (1/mm)	v (mm^2/s)	ρ (g/cm^3)
value	0.15	0.5	5	20	1.5	100	10	0.89

Table 5.3 Thermal physical parameters of AISI4130 alloy steel [4]

Material	Thermal conductivity K (W/m K)	Specific heat C_p (J/kg K)	Density (kg/m^3)
AISI 4130	42.7	477	7850

Table 5.4 Johnson–Cook parameters of AISI4130 alloy steel [4]

Material	A (MPa)	B (MPa)	C	m	n	T_m (°C)
AISI 4130	610	750	0.008	1	0.25	1432

Table 5.5 Specifications of the tool material

Material	Thermal conductivity κ (w/mK)	Specific heat C_p (J/kg K)	Density (kg/m^3)	Elasticity modulus (MPa)	Poisson's ratio
Uncoated carbide alloy	44	750	4370.1	560	0.3

5.2.2 Validation of Cutting Force

(1) Dry condition

Table 5.6 shows the cutting parameters and measured results under dry cutting conditions. The measured cutting force of Test No. 1 is shown in Fig. 5.12. Figure 5.12a shows the original signal of the cutting force, and Fig. 5.12b shows the filtered signal of the cutting force. From Fig. 5.12, is can be seen that the measured cutting force tends to be stable, and the force component F_r in the radial direction is approximately 0, which can be ignored. This is because the ratio of workpiece material wall thickness to outer diameter is 1:15, so the cutting process can be approximately considered as orthogonal cutting, which is consistent with the prediction model.

From Table 5.6, it is found that when other cutting conditions are the same, the cutting forces both in cutting direction and tangential direction increase with the increase of cutting depth, and decrease with the increase of cutting speed, but the trend is not very obvious. This is because when the cutting depth increases, the cutting area increases, the total amount of elastic–plastic deformation in the cutting process increases, so the cutting force increases. The cutting force decreases with

Table 5.6 Measured results of AISI 4130 alloy steel under dry cutting condition

Test No.	V (m/s)	DOC (mm)	Width of cut (mm)	Rake angle (deg)	Measured F_c (N)	Measured F_t (N)
1	1.049	0.0508	4.775	5	682.83	491.34
2	1.049	0.1016	4.775	5	1088.30	669.77
3	1.049	0.1524	4.775	5	1463.80	821.86
4	3.147	0.0508	4.775	5	632.68	501.03
5	3.147	0.1016	4.775	5	1059.50	697.31
6	3.147	0.1524	4.775	5	1394.10	758.45
7	2.098	0.0508	4.775	5	708.84	576.06
8	2.098	0.1524	4.775	5	1378.80	753.03

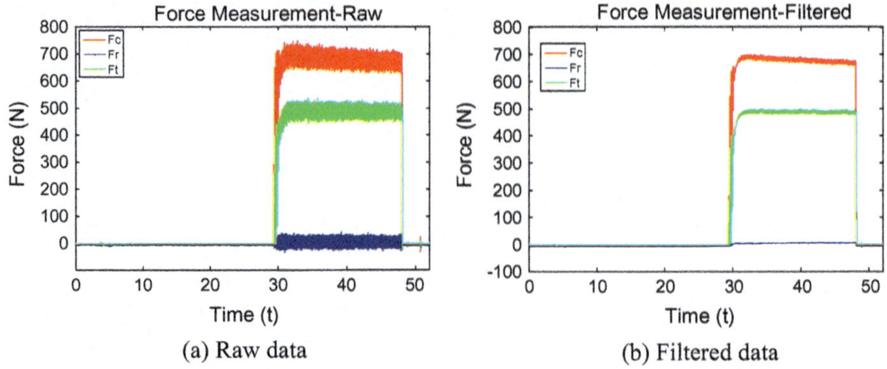

Fig. 5.12 Measured cutting forces for Test No. 1

the increase of the cutting speed, this is because when the cutting speed increases, the heat generated in the cutting process is too late to disperse, which leads to the increase of the temperature of the cutting area, and the softening of the material, so the cutting force is reduced.

Table 5.7 shows the flow stress, shear angle, friction coefficient, total cutting force and ploughing force predicted by the prediction model. The predicted results show that when the cutting speed is constant, the friction coefficient decreases with the increase of the feed rate, but the trend is not obvious. This is because when the feed rate increases, the normal stress on the front tool surface also increases, so the friction coefficient decreases slightly. In the cutting direction, both chip formation force and ploughing force increase with the increase of feed rate. Due to the increase in feed rate, that is, the cutting thickness increases, so the cutting force increases. However, as the feed rate increases, the chip deformation coefficient decreases, the friction coefficient also decreases, so the friction decreases. Due to the positive and negative effects, the increasing speed of cutting force is not proportional to the increasing speed of feed rate.

Table 5.7 Detailed prediction results for AISI 4130 alloy steel under dry condition

Test No.	k_{ab} (MPa)	Shear angle (deg)	μ	Pred F_c (N)	Pred F_t (N)	Pred P_c (N)	Pred P_t (N)
1	607.60	27.5	0.7118	501.17	271.47	101.40	240.75
2	594.36	27.5	0.7118	980.49	531.11	99.19	235.51
3	586.89	30.0	0.6087	1334.70	603.43	97.94	232.55
4	586.89	30.0	0.6087	444.90	201.14	97.94	232.55
5	578.85	30.0	0.6087	877.61	396.78	96.60	229.36
6	578.03	32.5	0.5111	1209.90	442.10	96.46	229.04
7	594.36	27.5	0.7118	490.25	265.55	99.19	235.51
8	578.85	30	0.6087	1316.40	595.17	96.60	229.36

When the feed rate is constant, the friction coefficient decreases with the increase of cutting speed, but the increasing trend is not obvious. When the cutting speed increases, the cutting force decreases significantly in the cutting direction. This is because the friction coefficient decreases after the cutting speed increases, so the deformation coefficient also decreases and the cutting force becomes smaller. On the other hand, the cutting speed increases, the cutting temperature also increases, the strength and hardness of the processed metal is reduced, leading to the reduction of cutting force. What's more, for chip formation force and ploughing force, cutting speed and feed rate have little effect on the cutting force which is perpendicular to the workpiece surface.

Under dry cutting condition, the decomposition results of cutting force prediction obtained according to the cutting force prediction model are shown in Fig. 5.13. Figure 5.13a represents the cutting force in the cutting direction, and Fig. 5.13b represents the cutting force in the thrust direction. Since the tool wear is not considered in the prediction model of cutting force in this book, the total cutting force mainly includes chip formation force due to shear deformation and plowing force due to friction between tool edge radius and the machined surface. From Fig. 5.13a, it is found that the cutting force in the cutting direction mainly comes from chip formation force. From Fig. 5.13b, it is found that chip formation force and plowing force are relatively balanced in the direction perpendicular to the cutting direction. The results show that the cutting force perpendicular to the cutting direction is significantly affected by the tool edge radius. This is because when the tool edge radius increases, the extrusion deformation between the tool tip and the machined surface increases during the cutting process, so the cutting force perpendicular to the machined surface increases.

Figure 5.14 shows the comparison between the measured force and the predicted cutting force under dry cutting condition. From Fig. 5.14, it is found that the predicted results are in good agreement with the experimental measurements. Within the explored range of experiments, the prediction error of cutting force in thrust direction

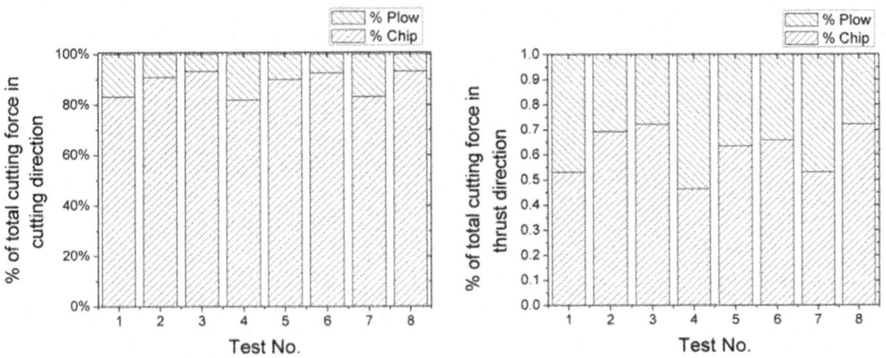

(a) In the cutting direction

(b) In the thrust direction

Fig. 5.13 Force breakdown under dry condition

is 0.79–16.84%, and the average prediction error is 7.82%. The prediction error of cutting force in thrust direction is 1.72–14.46%, and the average prediction error is 9.77%. By comparing the measured and predicted results, it is found that the predicted results are smaller than the measured results, which may be caused by the tool wear, which is not considered in the proposed model in this book. While in actual machining, the tool wear is inevitable.

(2) MQL condition

In the experiment, the MQL parameters mainly consider the cutting fluid flow rate and air-oil mixture ratio, and the air pressure remains unchanged during the experiment. The air-oil mixture ratio under MQL condition has three levels, they are high, medium and low, respectively. Firstly, the cutting force under the condition that the air-oil mixture ratio with medium level and the flow rate of lubricating fluid with 16 mL/h is discussed. The cutting parameters and measurement results are shown in Table 5.8.

According to the measured results in Table 5.8, under MQL condition, when other cutting parameters remain unchanged, the cutting force increases with the increase of cutting depth and decreases with the increase of cutting speed, and the decreasing trend is not obvious. This is because when the cutting depth increases, the cutting area increases, resulting in an increase in plastic deformation, which leads to an increase in cutting force. When the cutting speed increases, the cutting temperature rises, and the material hardness and strength decrease, resulting in the reduction of cutting force.

It is assumed that the flow rate of MQL is proportional to the thickness of boundary lubricating film, and its proportional coefficient is calibrated by the measured results. The measured data of Test No. 9 is randomly selected to calibrate the thickness of the boundary lubrication film and the correction coefficient of the heat transfer coefficient in the temperature model as Eq. (2.25). According to the comparison between the

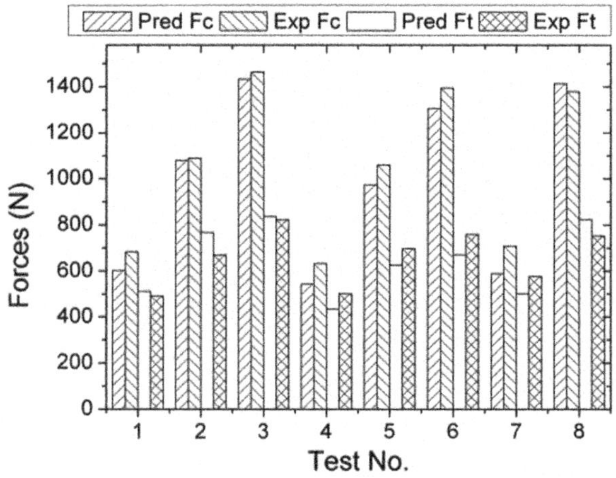

Fig. 5.14 Comparisons of cutting forces for AISI 4130 alloy under dry condition

Table 5.8 Measured cutting forces for AISI 4130 alloy steel under MQL condition

Test No.	V (m/s)	DOC (mm)	Flow rate (ml/h)	Air-oil mixture ratio mixture ratio	Measured F_c (N)	Measured F_t (N)
9	1.049	0.0508	16	M	676.07	485.77
11	1.049	0.1016	16	M	1075.50	659.93
12	1.049	0.1524	16	M	1484.10	851.08
13	2.098	0.0508	16	M	660.60	507.29
14	2.098	0.1016	16	M	1044.80	657.29
16	2.098	0.1524	16	M	1376.20	737.92
17	3.147	0.0508	16	M	642.56	512.70
18	3.147	0.1016	16	M	1057.00	710.36
19	3.147	0.1524	16	M	1375.70	748.22

measured results and the predicted results, when the flow rate of MQL is 16 mL/h, the boundary lubrication film thickness t_b is 0.1 μm. According to the analysis in Chap. 2, the air-oil mixture ratio mainly affects the heat transfer coefficient. In this experiment, when the air-oil mixture ratio is medium, the correction coefficient of heat transfer coefficient is 427. Table 5.9 shows the predicted friction coefficient, flow stress, shear angle, chip formation force and plowing force by the prediction model. The cutting forces of Test No. 10 and Test No. 15 were not measured. From Table 5.9, it can be seen that when the cutting speed is constant, the friction coefficient decreases with the increase of the feed rate, but the trend is not obvious. The cutting force in the thrust direction increases with the increase of feed rate. On the contrary, the cutting force in the thrust direction decreases significantly with increasing cutting speed. This is mainly because as the cutting speed increases, the cutting temperature also increases, the strength and hardness of the processed metal is reduced, resulting in a reduction in cutting force. For the plowing forces, both in cutting direction and thrust direction are insensitive to the change of cutting speed and feed rate.

For MQL machining, the prediction decomposition results of cutting force predicted by the proposed model are shown in Fig. 5.15. Figure 5.15a represents cutting force in the cutting direction and Fig. 5.15b represents cutting force in the thrust direction. From Fig. 5.15, it is found that in the cutting direction, the cutting force mainly comes from chip formation force, nearly 90% of the total cutting force. In the thrust direction, the chip forming force accounts for about 66% of the total cutting force. Compared with the decomposition results of cutting force under dry cutting condition, the effect of MQL on the percentage of chip formation force and plowing force in total cutting force is not obvious.

On the other hand, from Fig. 5.15 it can be seen that, the percentage of chip formation force in the total cutting force increases with the increase of feed rate. This is because the feed rate increases, the cutting work increases, the total cutting force increases as well. The increased cutting force mainly comes from the chip

Table 5.9 Detailed prediction results for AISI 4130 alloy steel under MQL condition

Test No.	k_{ab} (MPa)	Shear angle (deg)	μ	Pred F_c (N)	Pred P_c (N)	Pred F_t (N)	Pred P_t (N)
9	529.60	42.5	0.7098	510.5	88.38	275.66	209.85
11	457.02	45.0	0.7098	963.91	76.27	520.47	181.09
12	480.68	47.5	0.6070	1381.8	80.22	622.71	190.46
13	451.47	45.0	0.7098	476.1	75.34	257.08	178.89
14	454.09	47.5	0.6070	870.26	75.78	392.17	179.93
16	443.09	45.0	0.6070	1173.9	74.94	528.99	175.57
17	475.05	47.5	0.6070	455.22	79.28	205.14	188.23
18	441.90	45.0	0.6070	780.48	73.75	351.71	175.1
19	472.36	47.5	0.5097	1145.3	78.83	417.06	187.17

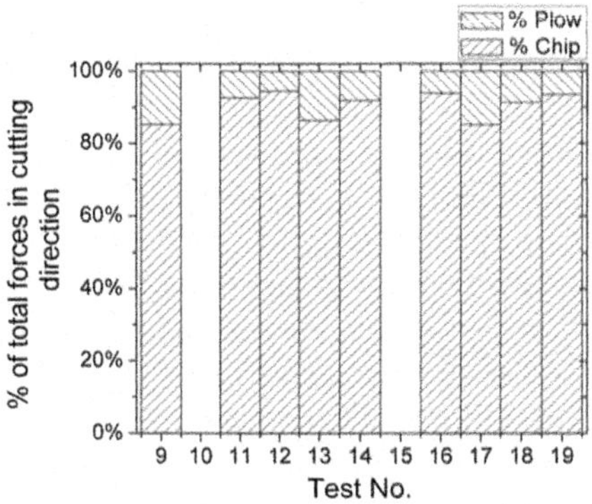

(a) In the cutting direction

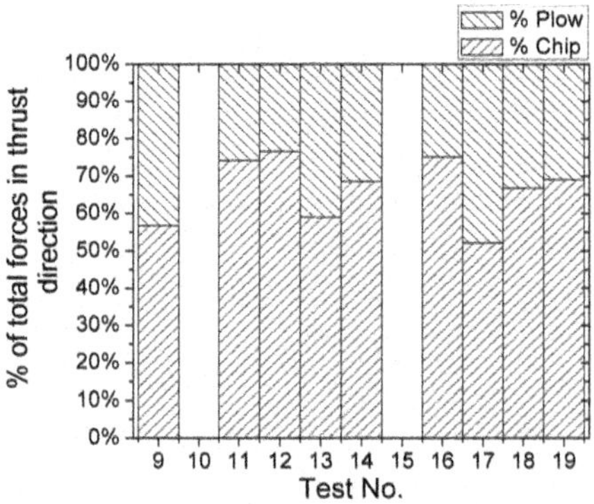

(b) In the thrust direction

Fig. 5.15 Force breakdown under MQL condition

formation force, and the plowing force mainly depends on the radius of the tool edge. Therefore, the increase of feed rate has little effect on the plowing force.

The comparison between the measured cutting force and the predicted results is shown in Fig. 5.16, when the flow rate of MQL is 16 mL /h and the oil-mist mixture ratio is medium. It can be seen that the predicted cutting forces are in good agreement with the measured results both in cutting direction and thrust direction. The prediction error of cutting force in cutting direction is 1.48–16.82% except for Test No. 18, and the average prediction error is 9.91%. The prediction error of cutting force in thrust direction is 0.06–23.27% except for Test No. 18, and the average prediction error is 10.61%. Comparing the measured with predicted values, it can be seen that the predicted cutting force is obviously smaller than the measured values, both in cutting direction and thrust direction, which is mainly caused by the failure of tool wear in the prediction model. Compared the measured data of Test No. 9 to Test No. 19, the prediction error has an increasing trend. This is because in the actual machining

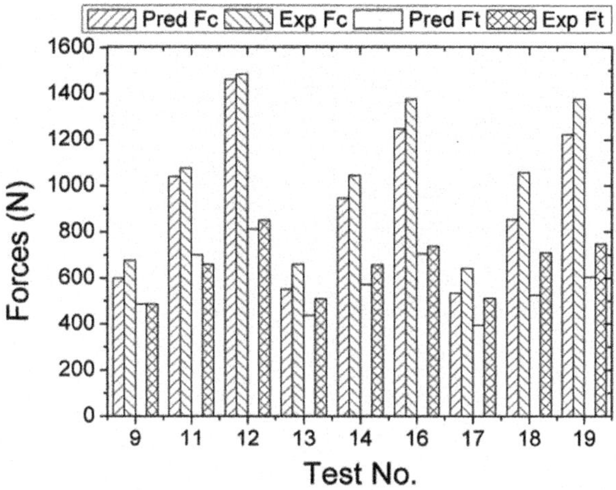

Fig. 5.16 Comparisons of cutting force for AISI 4130 alloy under MQL condition

process, the same tool was used from Test No. 9 to Test. No. 19. As the experiment went on, the tool wear became worse, so the prediction error caused by tool wear also increased.

When the cutting fluid flow is 32 ml/h, 48 ml/h and 64 ml/h respectively, and the air-oil mixture ratio is medium, the measured cutting force and predicted flow stress, shear angle, friction coefficient and cutting force are shown in Table 5.10. The measured data of Test No. 20 was randomly selected to calibrate the thickness of the boundary lubrication film t_b, and the measured results were compared with the predicted results. When the cutting fluid flow rate was 48 ml/h, the thickness of the boundary lubrication film t_b was 0.3 μm. While for Test No. 9, the flow rate of MQL is 16 ml/h, and the boundary lubrication film thickness t_b is 0.1 μm. It is assumed that the thickness of the boundary lubrication film t_b is proportional to the flow rate of MQL. Therefore, when the cutting fluid flow rate is 32 ml/h and 64 ml/h, the corresponding of the lubrication film thickness is 0.2 μm and 0.4 μm, respectively. According to the analysis in Sect. 2.3, it can be seen that the air-oil mixture ratio mainly affects the heat transfer coefficient in the cutting process. According to the comparison of the predicted results to the measured results of Test No. 9, the correction coefficient of heat convection coefficient λ is 427 when the air-oil mixture ratio is at medium level.

From Table 5.10, it shows that within the explored experimental range, when cutting parameters remain unchanged, the cutting forces both in cutting direction and thrust direction decrease first and then increase with the increase of cutting fluid flow rate, but the change trend is not obvious. For example, when the cutting speed is 2.098 m/s, the cutting depth is 0.0508 mm, and the flow rate is 32 ml/h, the measured cutting force is 672 N. When the flow rate increases to 48 ml/h, the measured cutting force is 669 N. The cutting force in thrust direction decreases with the increase of the cutting fluid flow rate. However, when the flow rate of MQL continues to increase to 64 mL /h, the measured cutting force in thrust direction is still 669 N, with no

Table 5.10 The measured and predicted cutting forces under different flow rates

Test No.	V (m/s)	DOC (mm)	Flow rate (ml/h)	k_{ab} (MPa)	Shear angle (deg)	μ	Predicted		Measured	
							F_c (N)	F_t (N)	F_c (N)	F_t (N)
20	2.098	0.0508	48	474	45	0.6521	530	409	669	521
21	2.098	0.1016	48	491	47.5	0.5585	944	546	1060	671
22	2.098	0.1524	48	446	50	0.5584	1357	700	1389	749
23	3.147	0.0508	48	502	50	0.5586	566	396	640	508
24	3.147	0.1016	48	445	50	0.5584	927	524	1026	662
25	3.147	0.1524	48	473	47.5	0.5070	1220	600	1350	717
26	2.098	0.0508	32	451	45	0.7078	550	434	672	525
27	2.098	0.1016	32	454	47.5	0.6053	944	570	1056	671
28	2.098	0.1524	32	443	45	0.6053	1245	702	1395	761
29	3.147	0.0508	32	475	47.5	0.6053	533	392	640	507
30	3.147	0.1016	32	442	45	0.6053	852	525	1075	709
31	3.147	0.1524	32	472	47.5	0.5084	1222	602	1398	775
32	2.098	0.0508	64	452	45	0.7038	547	431	669	515
33	2.098	0.1016	64	454	47.5	0.6020	939	565	1058	669
34	3.147	0.0508	64	475	47.5	0.6020	531	390	643	509
35	3.147	0.1016	64	442	45	0.6020	849	521	1032	665
36	3.147	0.1524	64	473	47.5	0.5056	1217	598	1346	719
37	2.098	0.1524	64	443	45	0.6020	1240	696	1389	749

obvious change. This is because in the actual cutting process, the continuous chip partly blocks the spray of MQL, thus affecting the lubrication and cooling effect of MQL. Therefore, the change of measured cutting force in the actual experiment is not obvious with the change of flow rate of MQL.

When the cutting speed is 2.098 m/s, the cutting depth is 0.0508 mm, and the air-oil mixture is at medium level, the comparison between the predicted cutting force and the measured cutting force under different flow rates is shown in Fig. 5.17a. It is found that the prediction errors of cutting force in cutting direction and thrust direction are 18.2%–20.6% and 16.3%–21.4% respectively. When the cutting speed is 3.147 m/s, the cutting depth is 0.1016 mm, and the air-oil mixture ratio is medium, the comparison between the predicted total cutting force and the measured total cutting force at different flow rates is shown in Fig. 5.17b. According to the comparison results, the prediction errors of cutting force in cutting direction and thrust direction are 9.6%–20.7% and 20.8%–25.9% respectively. It can be seen that within the explored experimental range, when other cutting conditions are the same, the cutting force decreases first and then increases with the increase of cutting fluid flow rate. According to the analysis of the penetration mechanism of cutting medium, when the cutting fluid flow rate increases, the lubrication effect is significantly improved, leading to the reduction of cutting force. However, when the cutting fluid flow rate increases to a certain extent, that is, when the optimal lubrication effect is achieved, the cutting force does not increase with the increase of the cutting fluid flow rate any more. In actual machining, the cutting force measured at different flow rates did not change significantly, which may be due to the continuous chip generated in the cutting process, resulting in the change of the position of the MQL nozzle, thus affecting the effect of MQL. Comparing Fig. 5.17a with Fig. 5.17b, it can be seen that the prediction error increases obviously, this is mainly because the prediction model did not consider the tool wear, and in actual machining process, from Test No. 20 to Test No. 37, the same tool is used. As the cutting time goes on, the effect of tool wear is more and more significant. In addition, it can be seen that the trend of increasing prediction error is particularly significant in the thrust direction. This is mainly because with the progress of cutting process, tool wear is intensified, according to the analysis of 3.1.2, tool wear will lead to the increase of extrusion friction between the tool tip and the machined surface, so the cutting force in the thrust direction increases, and the plowing force also increases.

The cutting forces measured under different air-oil mixture ratios are analyzed below. Table 5.11 shows the experimental measurements under different cutting parameters and MQL parameters, as well as the predicted flow stress, shear angle, heat transfer coefficient and cutting force. The comparison of predicted cutting forces and measured results is shown in Fig. 5.18. The heat transfer coefficient λ of low air-oil mixture ratio and high air-oil mixture ratio are calibrated by Test No. 38 and Test No. 42 respectively, and they are determined to be 512 and 342 according to the comparison between the measured value and the predicted value. By comparing the measured results with the predicted results, the prediction errors of cutting force in cutting direction and thrust direction are 10.3%–25.1% and 13.8%–22% respectively. Comparing the experimental results with the predicted results, it is found that the

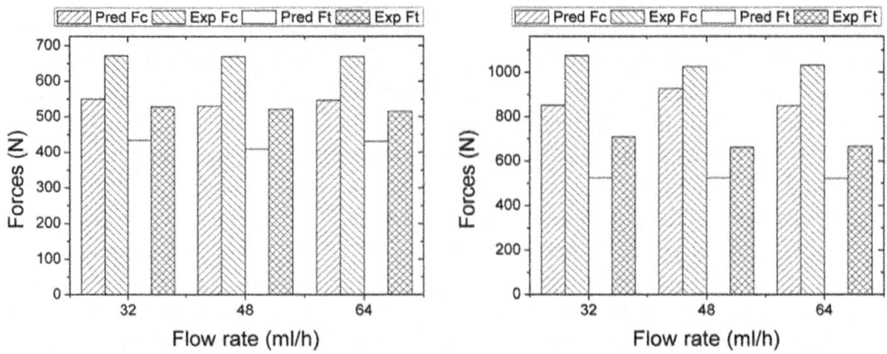

(a) V=2.098 m/s, DOC=0.0508 mm (b) V=3.147 m/s, DOC=0.1016 mm

Fig. 5.17 Comparisons of cutting force under different flow rates

predicted cutting forces are generally small. The prediction errors may come from: (1) the prediction model in this book is based on sharp tools, without considering the tool wear, which is inevitable in the actual machining; (2) In this experiment, the same tool was used in the experiments from Test No. 38 to Test No. 45. With the increase of tool wear, the prediction error also increased.

When the cutting speed is 2.098 m/s, the cutting depths are 0.0508 mm and 0.1016 mm, and the flow rates of MQL are 16 ml/h and 48 ml/h, the measured cutting forces under three air-oil mixture ratio levels are shown in Fig. 5.19. It can be seen that within the explored experimental range, when other cutting conditions remain unchanged, the cutting force decreases with the increase of air-oil mixture ratio. Under the condition of small cutting depth, the cutting force in thrust direction has an obvious downward trend, while the cutting force in cutting direction does not change significantly. This is because when the air-oil mixture ratio is increased, the lubrication and cooling effect of MQL is improved, so the cutting force is reduced. When the cutting depth increases, the cutting force increases, and the influence of the air-oil mixture ratio on the cutting force is less than that of the cutting depth. Therefore, the cutting force has no regular change in Fig. 5.19b.

(3) Traditional flood cooling condition

Table 5.12 shows the measured results and predicted results under traditional flood cooling condition.

According to the measured results, the curves of cutting force changing with cutting speed and cutting depth are shown in Fig. 5.20. It can be seen that under the traditional flood cooling condition, the cutting force both in cutting direction and thrust direction decreases with the increase of cutting speed, and increase with the increase of cutting depth. This is because when the cutting depth increases, the cutting area increases, and the plastic deformation in the cutting process intensifies, so the cutting force increases. When the cutting speed increases, the cutting temperature

Table 5.11 The cutting force under different flow rates and air-oil mixture ratios

Test No.	V (m/s)	DOC (mm)	Flow rate (ml/h)	Air-oil ratio	k_{ab} (MPa)	Shear angle (deg)	μ	h_{eff} (J/m^2 S °C)	Predicted		Measured	
									F_c (N)	F_t (N)	F_c (N)	F_t (N)
38	2.098	0.0508	48	L	458	45	0.7058	9.4121 e5	556	439	673	525
39	2.098	0.1016	48	L	458	47.5	0.6037	9.4121 e5	948	572	1057	664
40	2.098	0.1016	16	L	457	47.5	0.6070	9.4121 e5	953	576	1063	677
41	2.098	0.0508	16	L	458	45	0.7098	9.4121 e5	560	442	675	525
42	2.098	0.0508	16	H	446	42.5	0.7098	6.2748 e5	504	408	671	517
43	2.098	0.1016	16	H	451	47.5	0.6070	6.2748 e5	939	568	1062	677
44	2.098	0.1016	48	H	451	47.5	0.6037	6.2748 e5	934	563	1056	663
45	2.098	0.0508	48	H	446	42.5	0.7058	6.2748 e5	502	406	671	521

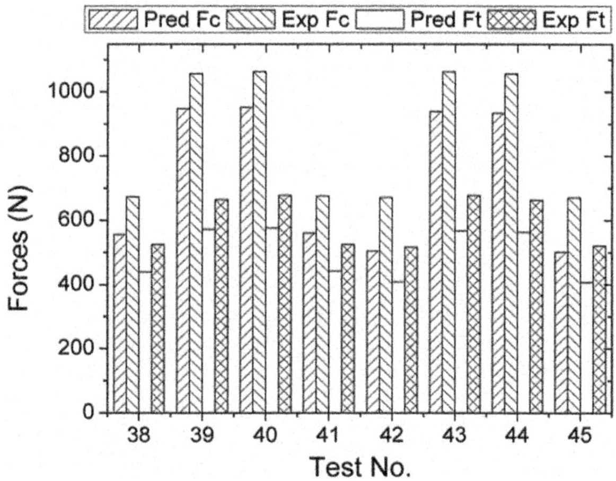

Fig. 5.18 Comparisons of cutting force under different flow rates and air-oil mixture ratios

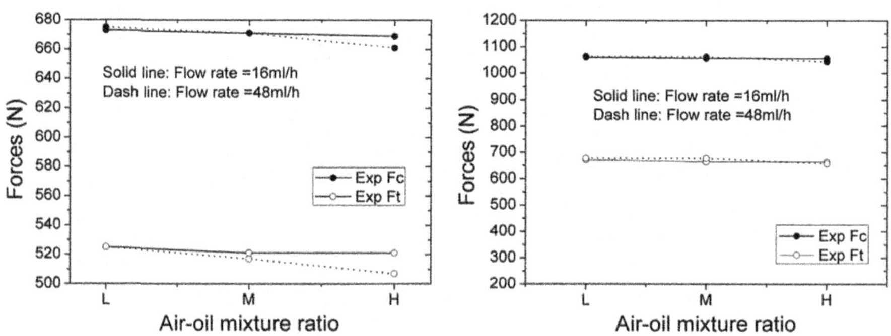

(a) V=2.098 m/s, DOC=0.0508 mm (b) V=2.098 m/s, DOC=0.1016 mm

Fig. 5.19 Relationship between measured cutting forces and flow rates and air-oil mixture ratios

increases, resulting in a decrease in the hardness and stiffness of the workpiece material, so the cutting force decreases.

The measured cutting force of Test No. 46 was used to calibrate the lubrication film thickness during the cutting process, and it was determined to be 1 mm and the correction factor of the heat transfer coefficient was 514 under the flood cooling condition by comparing the measured results with the predicted results. The comparison between the predicted cutting force and the measured results is shown in Fig. 5.21. It can be seen that the prediction error of cutting force in cutting direction is 4%–22%, and the average prediction error is 9.8%. While the prediction error of cutting force in thrust direction is 1.1%–24.5%, and the average prediction error is 6.7%. It is also found that the predicted cutting forces are significantly smaller than the experimental measurements, the reason is similar as dry and MQL condition. This is because the prediction model does not consider the effect of tool wear. Comparing

Table 5.12 The measured and predicted results under conventional flood cooling machining

Test No.	V (m/s)	DOC (mm)	k_{ab} (MPa)	Shear angle (deg)	μ	Predicted		Measured	
						F_c (N)	F_t (N)	F_c (N)	F_t (N)
46	2.098	0.0508	453	45	0.6919	538	422	667	518
47	2.098	0.1016	455	47.5	0.5919	924	551	1072	672
48	2.098	0.1524	444	47.5	0.5919	1315	718	1399	739
49	1.049	0.0508	538	45	0.6920	639	501	690	527
50	1.049	0.1016	458	45	0.6919	1012	672	1095	656
51	1.049	0.1524	492	47.5	0.5920	1457	796	1467	787
52	3.147	0.0508	487	47.5	0.5920	535	391	614	376
53	3.147	0.1016	443	47.5	0.5919	899	536	1030	642
54	3.147	0.1524	479	47.5	0.4973	1220	592	1354	703

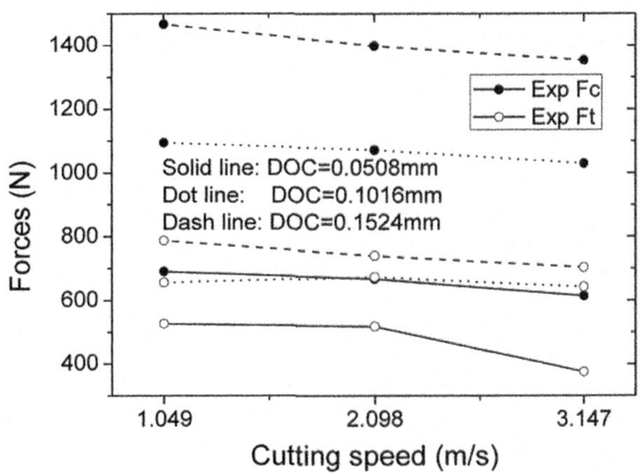

Fig. 5.20 Relationship between measured cutting forces and cutting parameters under conventional flood cooling machining

the prediction error of Test No. 46 and Test No. 54, it can be found that the prediction error increases with the increase of the test serial number, this is because the same tool is used in the cutting experiment. The tool wear increases with the cutting process goes on. Therefore, the prediction error caused by the tool wear increases.

In summary, for the cutting force prediction model, whether it is dry cutting, MQL or traditional flood cooling condition, the predicted cutting force is smaller than the measured result. However, the predicted results show good agreement with the measured results.

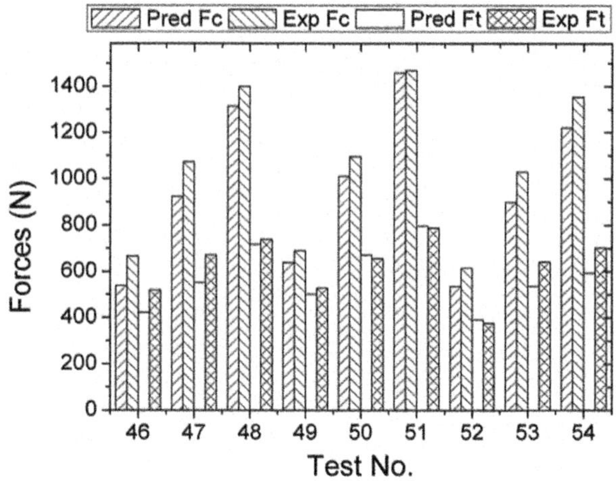

Fig. 5.21 Comparisons of cutting forces for AISI 4130 alloy under flood cooling machining

5.2.3 Validation of Cutting Temperature

(1) Dry condition

The predicted temperatures obtained under dry cutting conditions are compared with the experimental measurements as shown in Fig. 5.22. It is found that the temperature measured by the thermocouple differs significantly from the predicted results and the temperature obtained by the thermal imager. According to the analysis of the actual cutting test process, it can be seen that the thermocouple is glued to the rear tool surface through the heat-conductive synthetic adhesive and it takes some time for the heat to be transferred from the tool tip to the thermocouple. So, the temperature measured by the thermocouple is generally small due to the short cutting time not reaching the stable stage of temperature measurement. It can be verified by several phenomena: (a) Fig. 5.23 shows the temperature measured by the actual thermocouple. It is found that the measured cutting temperature rises and falls rapidly, without a stable region. It can be inferred that the actual cutting time was too short to cause the temperature measurement to reach a steady state, so the thermocouple measured a small temperature value. (b) During the actual test measurement, the cutting length was kept constant at 20 mm and when the cutting length was certain, the smaller cutting depth led to a longer cutting time. So, it was closer to the stable cutting condition and the measured temperature was closer to the actual temperature and it was found that the prediction error increased with the increase of cutting depth when comparing the thermocouple measured results of Tests No. 1, 2 and 3 with the prediction results. In addition, it is found in Fig. 5.22 that the temperature measured using the thermal imager in the third set of experimental conditions reached 1000 °C. By observing the experimental process recorded by the thermal imager, it is found that the continuous chips generated during the machining process blocked the position of the thermocouple, so the temperature recorded by the thermal imager was the temperature of the chips. A comparison of the predicted temperature and the temperature measured by the thermal imager revealed that the prediction error ranged from 2.3 to 15.5% with an average prediction error of 5.4%. Therefore, within the present experimental conditions, the predicted temperature obtained under dry cutting conditions is in better agreement with the cutting temperature measured by the thermal imager.

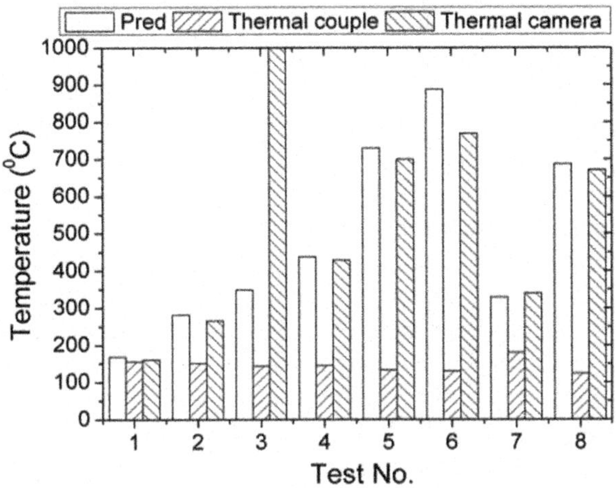

Fig. 5.22 Comparisons of cutting temperature under dry machining

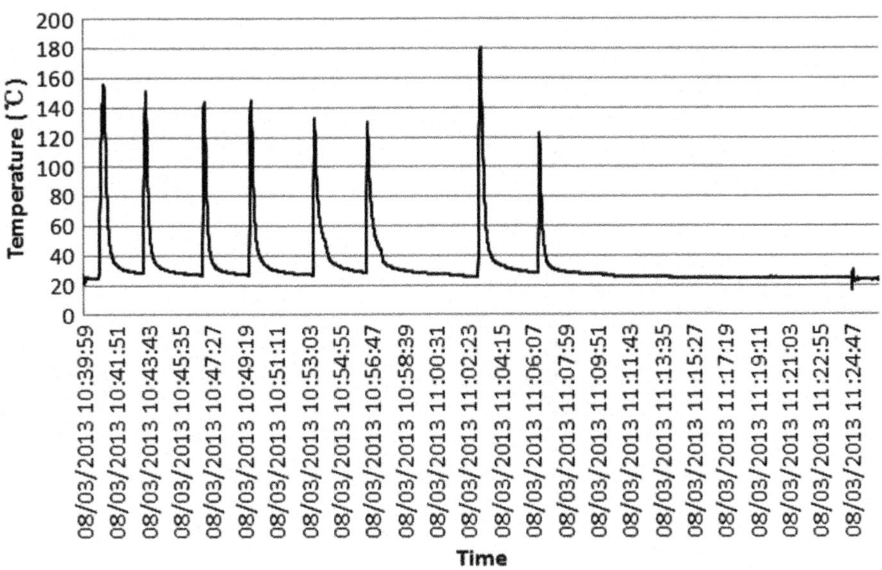

Fig. 5.23 The cutting temperature measured by thermal couple under dry machining

(2) MQL cutting process

When the flow rate of MQL is 16 ml/h and the air-oil mixture ratio is at medium level, the predicted temperature and the temperature measured by the thermal imager are compared as shown in Fig. 5.24. The temperature prediction error is 0.3–8.6% and the temperature prediction error of Test No. 11 is 21.0%, which may be the operation error in the actual measurement process. Overall, the model predictions are in better agreement with the actual measurements. As can be seen in Fig. 5.24, the cutting temperature increases with the increase of cutting speed and increases with the increase of depth of cut. This is because when the cutting speed increases, the amount of metal cut per unit time increases, so the cutting temperature increases. When the feed rate increases, the cutting force increases and the cutting temperature also increases, the contact length between the tool and the chip increases when the feed rate increases, the heat transfer area increases as well. Therefore, the increase of cutting temperature with feed rate is less than the increase of feed rate.

When the cutting fluid flow rate is 32 ml/h and 48 ml/h respectively and the oil mist mixing ratio is at medium level. The predicted temperature and the temperature measured by thermal imager are shown in Fig. 5.25. From the comparison results, it can be seen that the temperature prediction error is 0.3%–11.8% in this test measurement range and the average prediction error is 6.5%. The model prediction is in good agreement with the actual measurement results.

(3) Conventional high flow lubrication cutting process

When high flow lubrication cutting process, considering the safety of the experimental operation. The operation door of the machine must be closed for the test and the current infrared thermal imager cannot be measured through the glass. So, the use of thermal imager to measure the cutting temperature is only applicable to dry cutting and MQL cutting process. Thermocouples are used to measure the cutting temperature under high flow lubrication cutting conditions. However, because the

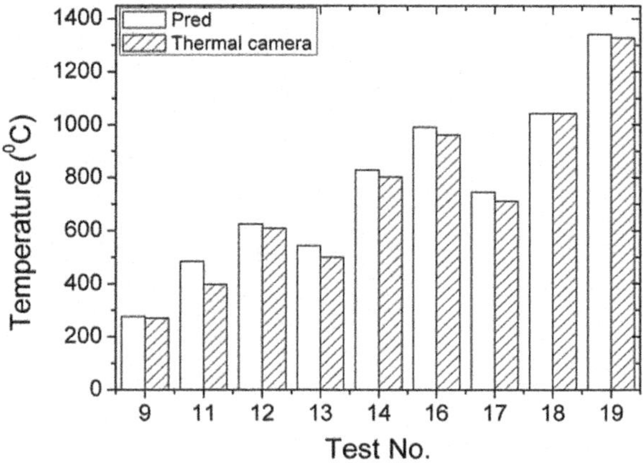

Fig. 5.24 Comparisons of cutting temperature under MQL machining

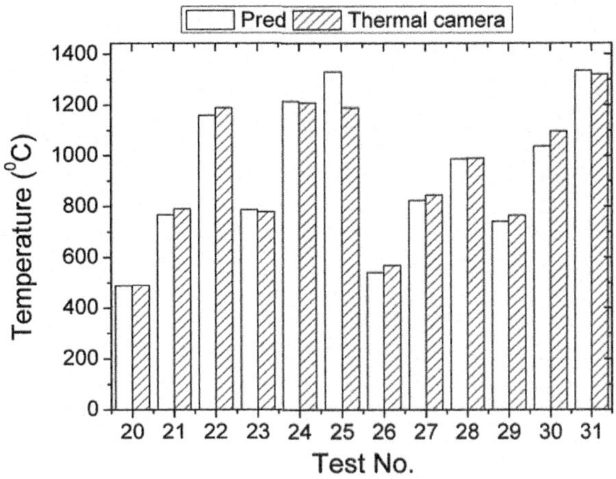

Fig. 5.25 Comparisons of cutting temperature under different flow rates and cutting parameters

actual cutting time is too short to reach the steady state of the measured temperature. The cutting temperature under conventional high-flow lubrication conditions is not discussed here.

5.2.4 Validation of Residual Stress Prediction Model

(1) Dry cutting processing

Test No. 1 was randomly selected to calibrate the algorithm constants ζ in the residual stress prediction model under dry cutting conditions. By comparing experimental measurements of average residual stresses in machined parts from the subsurface to a depth of 100 μm with predicted results. The value ζ of 0.15 was fitted using the least squares method. Figure 5.26 shows the predicted results of tangential residual stresses and radial residual stresses for Test No. 1 compared with the experimental measurements. As can be seen from Fig. 5.26a, for the tangential residual stress. The maximum residual compressive stress value predicted from the model is −178 MPa and the depth is 51 μm. The actual measured maximum residual compressive stress is −112 MPa and the depth is 52 μm. Both the model prediction results and the actual measurement results show that the residual stress basically tends to 0 at a depth of 100 μm. As can be seen from Fig. 5.26b, for the radial residual stresses. Both the magnitude and depth of the maximum residual compressive stress, the theoretical predictions are in good agreement with the actual measurement results. It can be seen that the predicted and measured residual stresses agree well in terms of distribution trends and orders of magnitude under the present experimental conditions.

Figure 5.27 shows the comparison between the predicted residual stresses and the actual measured results for Test No. 2. As seen in Fig. 5.27a, for the tangential residual stress. The predicted maximum residual compressive stress obtained is −234.8 MPa

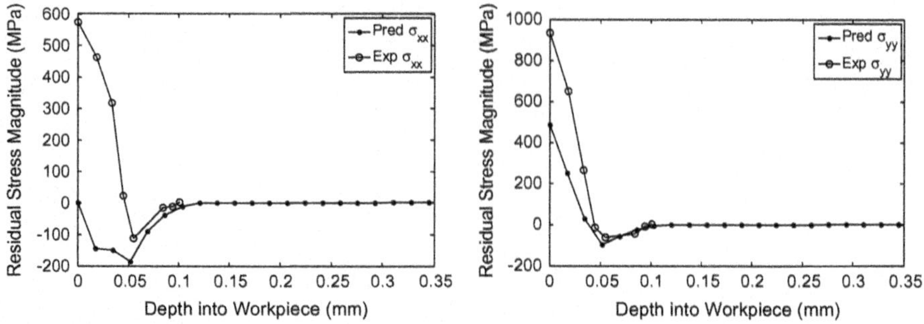

(a) Tangential residual stress (b) Radial residual stress

Fig. 5.26 Comparisons of predicted and measured residual stresses for Test No. 1

at a depth of 69.1 μm. The maximum residual compressive stress measured in the test was −102.3 MPa at a depth of 85.1 μm. The predicted residual stress is numerically greater than the test measurement. As can be seen in Fig. 5.27b, for the radial residual stress. The predicted maximum residual compressive stress obtained is −123.3 MPa at a depth of 86.3 μm. The maximum residual compressive stress measured in the test was −80.1 MPa at a depth of 86.3 μm. The predicted results were in good agreement with the actual measured values.

Figure 5.28 shows the comparison between the predicted residual stresses and the actual measured results for Test No. 3. As seen in Fig. 5.28a, for the tangential residual stress. The predicted maximum residual compressive stress obtained is −234.7 MPa at a depth of 86.3 μm. The maximum residual compressive stress measured in the test was −140.3 MPa at a depth of 79.9 μm. The predicted residual stress was numerically greater than the test measurement. As can be seen in Fig. 5.28b, for the radial residual stress. The predicted maximum residual compressive stress obtained is −137.7 MPa at a depth of 86.3 μm. The maximum residual compressive stress measured in the test was −101.6 MPa at a depth of 79.9 μm. Comparing the radial

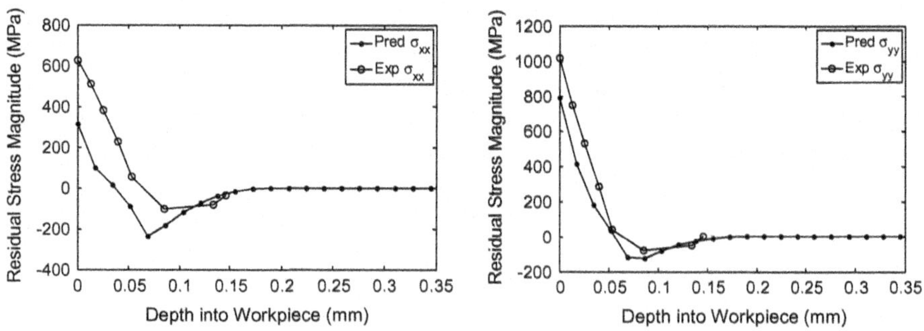

(a) Tangential residual stress (b) Radial residual stress

Fig. 5.27 Comparisons of predicted and measured residual stresses for Test No. 2

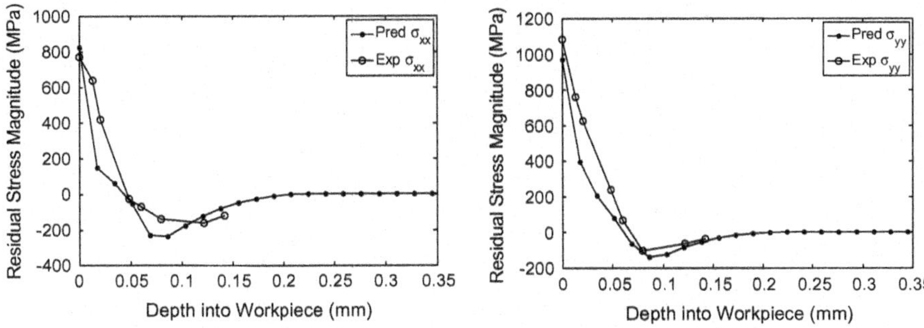

(a) Tangential residual stress (b) Radial residual stress

Fig. 5.28 Comparisons of predicted and measured residual stresses for Test No. 3

residual stresses and tangential residual stresses in Group 3, it was found that the radial residual stresses coincided better than the tangential residual stresses.

Comparing Figs. 5.26, 5.27 and 5.28, it is found that the residual stresses at the surface of the workpiece (depth is 0), both radial and tangential residual stresses, increase with increasing depth of cut for both the model predictions and the experimental measurements. This is because when the depth of cut increases, the cutting force and cutting temperature also increase. So, the stress on the surface of the workpiece due to mechanical and thermal loads also increases, which leads to an increase in residual stress on the surface of the workpiece.

A comparison of the predicted residual stresses and the actual measured results for Test No. 4 is shown in Fig. 5.29. For the tangential residual stress, the predicted maximum residual compressive stress is −260.0 MPa at a depth of 34.6 μm. The maximum residual compressive stress measured in the test was −88.5 MPa at a depth of 62.88 μm. For the radial residual stress, the predicted maximum residual compressive stress is −134.8 MPa at a depth of 51.8 μm. The maximum residual compressive stress measured in the test was −134.5 MPa at a depth of 62.88 μm. From the comparison results, it can be seen that the predicted maximum residual compressive stress depth is smaller than the actual measured maximum residual stress depth for both radial and tangential residual stresses. This may be caused by the small cutting force obtained from the model prediction. Because the residual stress is predicted based on the cutting force and cutting temperature, the cutting force during the actual test is greater than the cutting force predicted by the model because the prediction model does not consider tool wear, which is inevitable during the actual test. Resulting in the depth of the actual maximum residual stress is greater than the predicted value of the model. It can be seen that the maximum residual stress is mainly influenced by the cutting force. In general, the predicted residual stresses and the experimental measurements are in good agreement in terms of magnitude and trend distribution. The large deviation between the predicted and measured residual stress models for the surface layer of the workpiece (depth is 0) may be due to the fact that the surface of the workpiece is sensitive to external environmental influences,

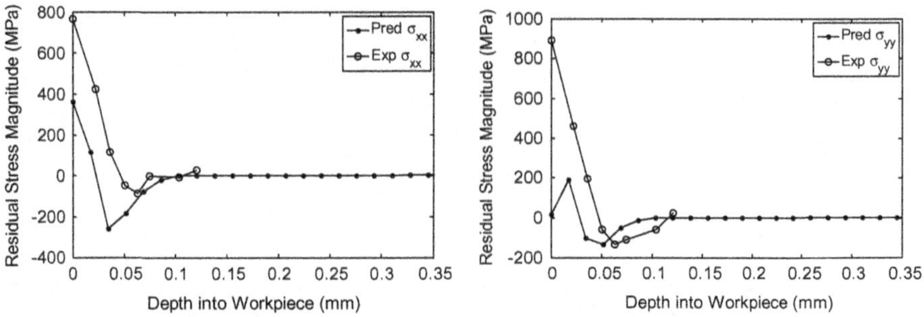

(a) Tangential residual stress (b) Radial residual stress

Fig. 5.29 Comparisons of predicted and measured residual stresses for Test No. 4

such as surface oxidation or corrosion, which can affect the change of residual stress in the surface layer.

(2) MQL cutting process

Test No. 9 is used to calibrate the algorithm constants ζ in the residual stress prediction model under MQL conditions. The average residual stresses measured and predicted by least-squares fitting of the test from the subsurface of the machined part to a depth of 100 μm, ζ is calibrated to be 0.2. Figure 5.30 shows the predicted results of tangential and radial residual stresses for Group 9 tests compared with the experimental measurements. For the tangential residual stress, the maximum residual compressive stress was −152.8 MPa at a depth of 24 μm. The actual measured maximum residual compressive stress was −52.8 MPa at a depth of 38.6 μm. The predicted residual stresses are numerically large relative to the experimental measurements, but the trends are more consistent. For the radial residual stress, the maximum residual compressive stress is −96.4 MPa at a depth of 36.3 μm. The maximum residual compressive stress measured in the test was −36.4 MPa at a depth of 38.65 μm. When the depth is 60 μm, the residual stress basically tends to 0. The predicted radial residual stresses match better than the tangential residual stresses.

Figure 5.31 shows the predicted tangential and radial residual stresses for Test No. 11 under MQL condition and the experimental measurements. As can be seen from the figure, for the tangential residual stress. The maximum residual compressive stress predicted is −144.4.0 MPa at a depth of 36.3 μm. The maximum residual compressive stress measured was −129.3 MPa at a depth of 40.2 μm. For the radial residual stress, the predicted maximum residual compressive stress is −180.2 MPa at a depth of 36.3 μm. The maximum residual compressive stress measured by the test was −77.4 MPa at a depth of 40.2 μm. The predicted values agree well with the actual measured values. Comparing Fig. 5.31a and b, it can be seen that the tangential residual stress is greater than the radial residual stress value and the depth of the residual stress affected layer in the tangential direction is deeper than that of the radial residual stress affected layer.

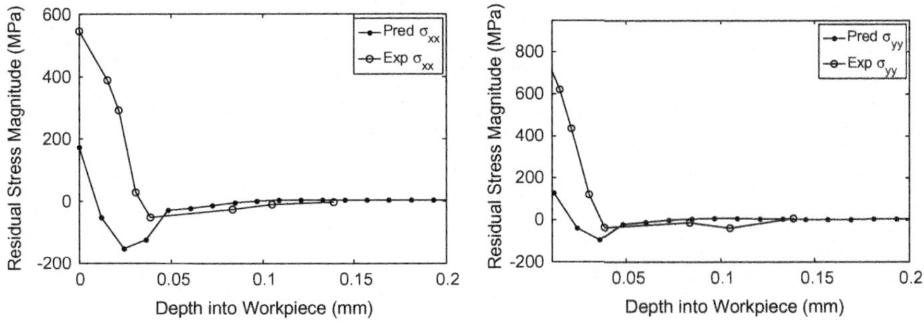

(a) Tangential residual stress (b) Radial residual stress

Fig. 5.30 Comparisons of predicted and measured residual stresses for Test No. 9

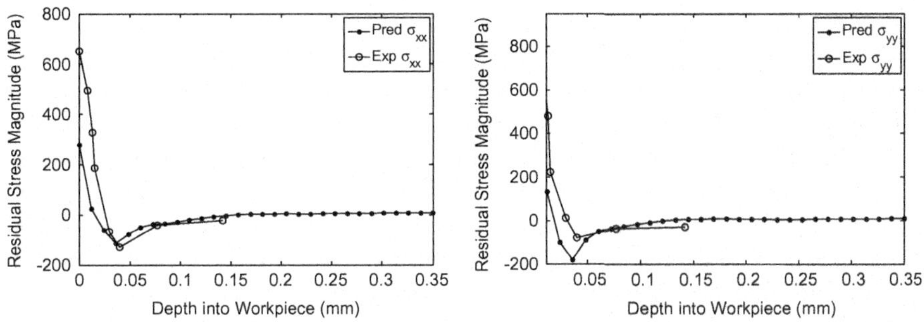

(a) Tangential residual stress (b) Radial residual stress

Fig. 5.31 Comparisons of predicted and measured residual stresses for Test No. 11

The predicted tangential and radial residual stresses for Test No. 12 under MQL condition are compared with the experimental measurements as shown in Fig. 5.32. As seen in Fig. 5.32a, for the tangential residual stress. The predicted maximum residual compressive stress obtained is −158.8 MPa at a depth of 48.4 μm. The maximum residual compressive stress measured was −176.0 MPa at a depth of 57.4 μm. As can be seen in Fig. 5.32b, for the radial residual stress. The predicted maximum residual compressive stress obtained is −185.8 MPa at a depth of 48.4 μm. The maximum residual compressive stress measured by the test was −93.4 MPa at a depth of 48.5 μm. The predicted values agree well with the actual measured values. Comparing Fig. 5.32a with Fig. 5.32b, it can be seen that for the surface layer of the workpiece (depth is 0) the tangential residual stresses are greater than the radial residual stress values. Overall, the predicted residual stresses agree well with the actual measurements in terms of values and distribution trends.

Comparing the experimental measurements in Figs. 5.30, 5.31 and 5.32, it can be seen that for the tangential residual stresses. The residual tensile stresses at the

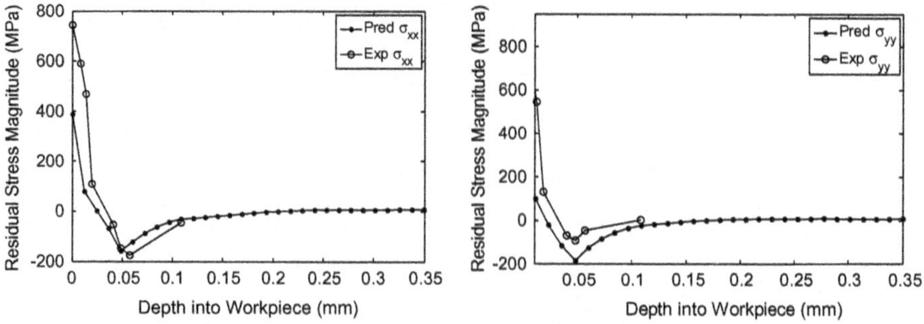

(a) Tangential residual stress (b) Radial residual stress

Fig. 5.32 Comparisons of predicted and measured residual stresses for Test No. 12

surface of the workpiece (depth is 0) as well as the maximum residual compressive stresses gradually increase. This is due to the fact that when the depth of cut increases, the cutting force and cutting temperature both increase and the stresses within the workpiece due to mechanical and thermal loads also increase, resulting in an increase in residual stresses on the workpiece surface. For the radial residual stress, the increasing trend is not obvious. According to the cutting force measurement results, it can be seen that when the depth of cut increases. The rate of increase of tangential force is greater than the rate of increase of axial force, thus it can be seen that the effect of depth of cut on tangential residual stress is more significant.

The predicted residual stresses for Test No. 19 under MQL conditions are shown in Fig. 5.33 along with the experimental measurements. As can be seen in Fig. 5.33, the results obtained from the model predictions are in good agreement with the experimental measurements, both for tangential and radial residual stresses.

Comparing Figs. 5.32 and 5.33, the cutting speed of Test No. 12 is 1.049 m/s and the cutting speed of Test No. 19 is 3.147 m/s and other cutting conditions are unchanged. The measured tangential force in Test No. 12 was 1484 N, the axial force

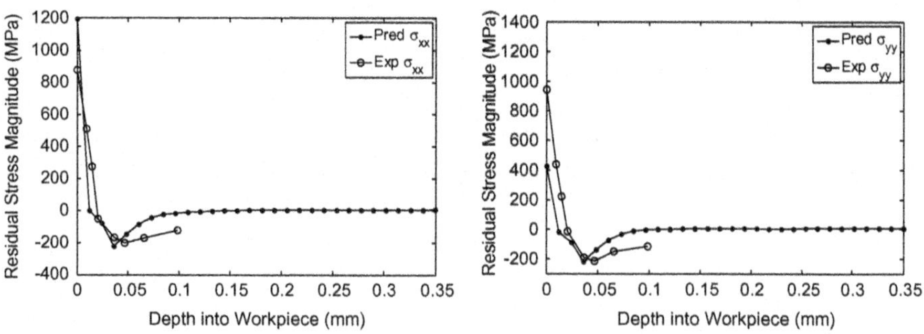

(a) Tangential residual stress (b) Radial residual stress

Fig. 5.33 Comparisons of predicted and measured residual stresses for Test No. 19

was 851 N and the measured cutting temperature was 609.8 °C. The tangential force measured in Test No. 19 was 1376 N, the axial force was 748 N and the measured cutting temperature was 1327.6 °C. According to the verification of cutting force and cutting temperature model, it is known that the cutting force decreases with the increase of cutting speed and the cutting temperature increases significantly with the increase of cutting speed. Comparing Figs. 5.32 and 5.33, it can be seen that the residual tensile stress in the surface layer of the workpiece (depth 0) increases significantly with the increase in cutting speed and the maximum residual compressive stress increases insignificantly. This is because the cutting temperature increases significantly when the cutting speed increases and the actual measured residual tensile stress in the surface layer of the workpiece also increases significantly. So, it can be inferred that the stress state in the surface layer of the workpiece is mainly affected by the cutting temperature. The trend of the maximum residual compressive stress is not obvious. So, it can be inferred that the cutting temperature has no significant effect on the maximum residual compressive stress and the residual compressive stress in the workpiece is mainly determined by the mechanical load.

The predicted residual stresses and experimental measurements for Test No. 36 under MQL condition are shown in Fig. 5.34. As seen in Fig. 5.34a, for the tangential residual stress. The predicted maximum residual compressive stress obtained is −130.5 MPa at a depth of 24.2 μm. The maximum residual compressive stress measured was −90.2 MPa at a depth of 37.7 μm. As can be seen in Fig. 5.34b, for the radial residual stress. The predicted maximum residual compressive stress obtained is −143.7 MPa at a depth of 24.2 μm. The maximum residual compressive stress measured was −84.2 MPa at a depth of 37.7 μm. Both radial and tangential residual stresses, the predicted results and the actual measured results are in better agreement.

Comparing Figs. 5.33 and 5.34, it can be seen that the residual tensile stress in the surface layer of the workpiece (depth is 0) is significantly reduced for both radial and tangential residual stresses and the trend of reduction of the maximum residual compressive stress in the workpiece is not obvious. This is because the parameters of Test No. 19 and Test No. 36 are unchanged except for the difference in the flow rate of

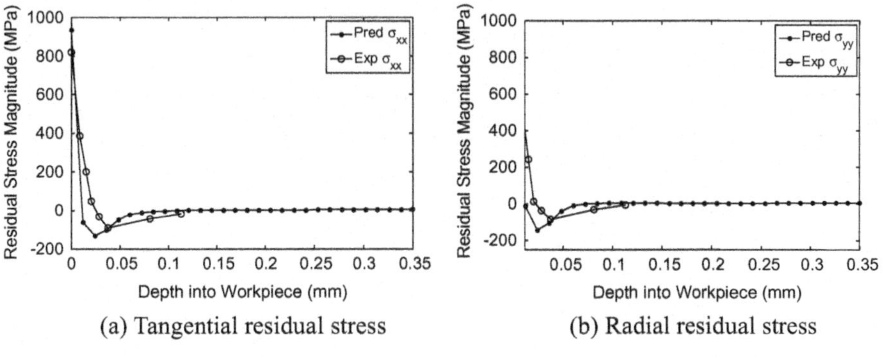

(a) Tangential residual stress (b) Radial residual stress

Fig. 5.34 Comparisons of predicted and measured residual stresses for Test No. 36

the MQL cutting fluid. From the test results of cutting force and cutting temperature, it can be seen that when the flow rate of MQL cutting fluid increases, the cutting temperature decreases significantly, the cutting force decreases more slowly and the residual tensile stress on the surface of the workpiece decreases. From this, it can be inferred that the stress state of the surface layer of the workpiece is mainly determined by the thermal load.

In summary, the residual stresses predicted by the MQL model agree well with the actual measured residual stresses in terms of values and distribution trends. From the analysis of the residual stress measurement results under MQL conditions, it is clear that the residual stress on the workpiece surface can be controlled by changing the MQL parameters。

(3) Conventional high flow lubrication cutting process

Figure 5.35 shows the predicted residual stresses and test measurements for Test No. 49 test under conventional high flow lubrication cutting conditions. As can be seen in Fig. 5.35, for the tangential residual stress. The predicted maximum residual compressive stress obtained is −169.6 MPa at a depth of 24.2 μm. The maximum residual compressive stress measured was −199.2 MPa at a depth of 42.9 μm. For the radial residual stress, except for the surface stress (−404.7 MPa). The predicted maximum residual compressive stress is −85.7 MPa at a depth of 36.3 μm. The maximum residual compressive stress measured by the test was −213.2 MPa at a depth of 42.9 μm. The predicted value of surface residual stress deviates greatly from the actual measured value and the reasons for the prediction error are mainly from: (1) The stress on the surface layer of the workpiece is sensitive to the influence of the external environment, such as surface oxidation or corrosion will affect the residual stress on the surface layer of the workpiece. (2) Based on the analysis of the cutting force measurement results, it is clear that the cutting force predicted according to the model deviates from the actual measurement results due to the severe wear of the tool under high flow lubrication conditions, thus affecting the accuracy of the residual stress prediction.

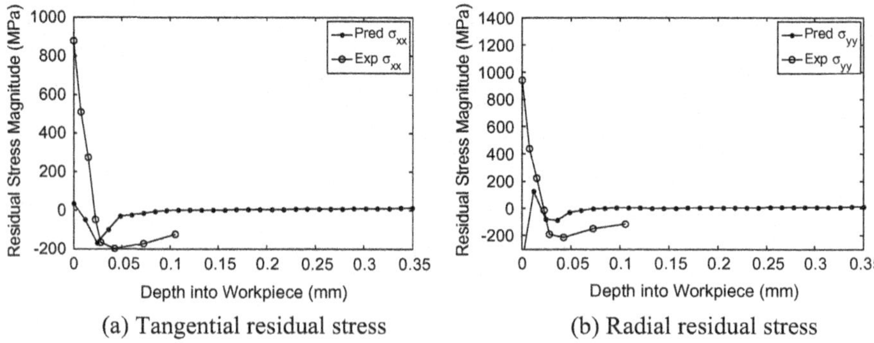

Fig. 5.35 Comparisons of predicted and measured residual stresses for Test No. 49

5.3 Summary

This chapter introduces the orthogonal turning test of AISI4130 alloy steel, measuring the cutting force, cutting temperature and residual stress on the surface of the machined part during the cutting process. And comparing the model prediction results with the actual measurement results, the following conclusions were obtained.

(1) Whether it is dry cutting, MQL or traditional high flow lubrication cutting process. The cutting force decreases with the increase of cutting speed and increases with the increase of feed. In this experimental range, the cutting force decreases with the increase of MQL cutting fluid flow rate and decreases with the increase of air to cutting fluid mixing ratio. The ratio decreases as the ratio increases. The predicted cutting force error was controlled within 30% for the three lubrication conditions.

(2) Comparing the cutting temperature measured by the thermal imager with the model predictions. It was found that for dry cutting and MQL cutting process. The results obtained from the temperature model prediction are in good agreement with the actual measurement results, and the prediction error is controlled within 20%. Due to the short cutting time, the temperature measured by the thermocouple did not reach a steady state, so the measurement results were generally small.

(3) Comparing the experimental measurements of tangential residual stress and radial residual stress. It was found that the tangential residual stress was greater than the radial residual stress and the depth of the layer affected by the tangential residual stress was deeper than that affected by the radial residual stress. The residual stresses in the surface layer of the workpiece (depth is 0) are mainly affected by the cutting temperature and are mostly expressed as residual tensile stresses. The maximum residual compressive stress in the workpiece is mainly influenced by the cutting force. The analysis of the predicted results and the actual measurement results shows that the residual stresses obtained by the model prediction agree well with the actual measured residual stresses in terms of values and distribution trends.

References

1. http://www.mcmaster.com.
2. Stachowiak G, Batchelor AW (2011) Engineering tribology. Butterworth-Heinemann
3. Kato S, Marui E, Hashimoto M (1998) Fundamental study on normal load dependency of friction characteristics in boundary lubrication. Tribol Trans 41(3):341–349
4. Shackelford JF, Alexander W, Park JS (2000) CRC materials science and engineering handbook, 3rd edn. CRC Press

Chapter 6
Sensitivity Analysis of Machined Residual Stress in MQL Machining

In order to quantitatively analyze how the machined residual stress varies with the cutting conditions in MQL machining, a sensitivity analysis of residual stress is carried out based on the verified MQL prediction model. The cutting conditions mainly include the MQL parameters, cutting parameters, and tool geometry parameters.

6.1 Selection of the Parameters

Based on the verified MQL prediction model, a sensitivity analysis of residual stress is introduced. From the introduction of the prediction model, it is known that the residual stress prediction model inputs mainly include the MQL parameter, cutting parameters, and tool geometry parameter.

From the analysis in Chap. 2, it is known that the MQL parameter involves many factors, such as cutting fluid characteristics, the flow rate of the cutting fluid, air-oil mixture ratio, air pressure, nozzle size, nozzle orientation and location. In this book, the MQL parameters mainly consider the flow rate of cutting fluid and air-oil mixture ratio, and other parameters are constants in the prediction model. Cutting parameters mainly include cutting speed, feed rate and the width of cut. The tool parameters mainly consider the tool rake angle and the tool edge radius. The influences of the MQL parameters, cutting parameters and tool parameters on cutting force, cutting temperature, and residual stress in MQL machining are emphatically discussed.

© The Author(s), under exclusive license to Springer Nature Singapore Pte Ltd. 2023

X. Ji, *Minimum Quantity Lubrication Machining*,
https://doi.org/10.1007/978-981-19-7087-0_6

6.2 Influence of MQL Parameters

6.2.1 Influence of Boundary Lubrication Film Thickness

From the introduction in Sect. 2.2.2, it is known that one effect of MQL is lubrication, which is reflected by the friction coefficient. While the friction coefficient in MQL machining largely depends on the boundary lubrication film thickness. It is assumed that the flow rate of MQL is proportional to the boundary lubrication film thickness. The boundary lubrication film thickness equals to zero under dry condition. Herein, the effect of flow rate of MQL is discussed in details. Take the lathe turning of AISI 4130 alloy steel with non-coated carbide tool as an example, the workpiece is with an outer diameter of 63.5 mm, a tube wall thickness of 4.775 mm, and a length of 60 mm, and its material properties are shown in Tables 5.3 and 5.4. The tool geometry parameters are shown in Table 5.5. The cutting speed is 1.5 m/s, the feed rate is 0.1 mm/r, the width of cut is 2 mm, and air-oil mixture ratio is at medium level. From the discussion in Sect. 5.2.2, it is known that when the flow rate of MQL is 16 mL/h, the boundary lubrication film thickness t_b equals to 0.1 μm. Since the flow rate of MQL is assumed to be proportional to the thickness of boundary lubricating film, here five different flow rates of MQL are selected, they are 0, 16 mL/h, 32 mL/h, 64 mL/h, 128 mL/h, respectively. The corresponding boundary lubrication film thicknesses are 0, 0.1 μm, 0.2 μm, 0.3 μm, and 0.4 μm, respectively. The predicted results of friction coefficient, average temperature at shear plane, average temperature at tool-chip interface, cutting forces, maximum residual stress and its depth, average residual stress are shown in Table 6.1. The maximum residual stress refers to the maximum compressive residual stress. Since the residual stress at the surface of the workpiece can be influenced by many factors, the average residual stress refers to the average value of the residual stress from the subsurface layer to the depth of 100 μm.

The relationship between the friction coefficient and the boundary lubrication film thickness is shown in Fig. 6.1. It can be seen that with the increase of boundary lubricating film thickness, the friction coefficient decreases linearly. This is because the thickness of the lubricating film increases, the lubricating effect increases significantly, and thus the coefficient of friction decreases. It also can be inferred that the friction coefficient decreases linearly with the increase of the flow rate.

Figure 6.2 shows the average temperature at the shear plane and the tool-chip interface under different boundary lubrication film thicknesses. It can be seen that with the increase of the thickness of the boundary lubricating film, the average temperature at the shear plane decreases at the beginning and then remains stable, while the average temperature at the tool-chip interface increases. This is because when the boundary lubrication film thickness increases, the friction coefficient decreases, and the friction heat decreases as well. Therefore, the average temperature at the shear plane has a slight decrease at the beginning. However, when the boundary lubrication film thickness increases to a certain extent, the boundary lubrication becomes fluid lubrication, and its lubrication effect is no longer obvious. Therefore, the average temperature at the shear plane decreases first and then tends to be stable. In terms

Table 6.1 Detailed predicted results under different lubrication film thickness

t_b (μ m)	μ	T_{AB} (°C)	T_{int} (°C)	F_c (N)	F_t (N)	Maximum value (MPa)			Average value (MPa)	
						σ_{xx}	σ_{yy}	Depth (μm)	σ_{xx}	σ_{yy}
0	0.6087	456	388	406	247	−107.9367	−131.5318	34.6	−41.1376	−36.3182
0.1	0.5920	444	394	399	239	−105.4908	−126.6421	34.6	−43.0263	−33.5976
0.2	0.5753	437	403	391	231	−105.6719	−116.7954	34.6	−44.0927	−29.9024
0.3	0.5587	434	414	382	222	−102.2790	−104.2238	34.6	−43.9904	−25.7818
0.4	0.5420	436	429	401	224	−109.1627	−109.0388	34.6	−39.2182	−32.6667

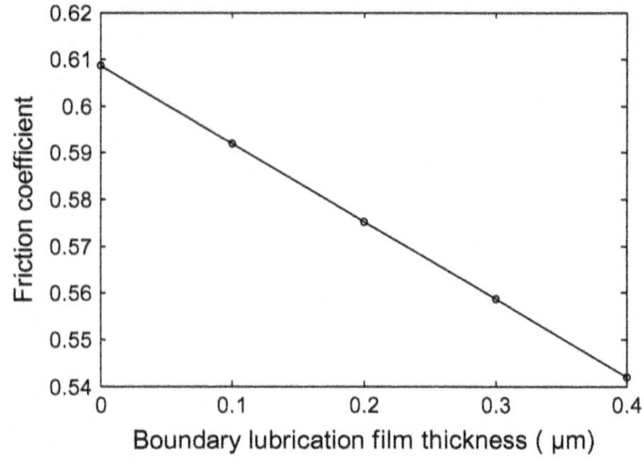

Fig. 6.1 Relationship between friction coefficient and boundary lubrication film thickness

of the average temperature at tool-chip interface, with the increase of the boundary lubrication film thickness, the lubrication effect is enhanced, but the cooling effect is relatively weakened, so the average temperature at tool-chip interface increases.

The relationship between cutting forces and boundary lubrication film thicknesses is shown in Fig. 6.3. It can be seen that both the cutting force and the thrust force have a slight decrease with the increase of the boundary lubrication film thickness. When the lubricating film increases to a certain value, the cutting force starts to increase. Compared to the thrust force, the change trend of the cutting force is more obvious. Therefore, in order to obtain the small cutting force, there is a certain optimization range for the boundary lubrication film thickness. This also explains that within a certain range of the flow rate, the cutting force under MQL condition is smaller than that under the conventional flood cooling condition.

The relationship between the residual stresses and boundary lubrication film thicknesses is shown in Fig. 6.4. It can be seen that the maximum radial compressive

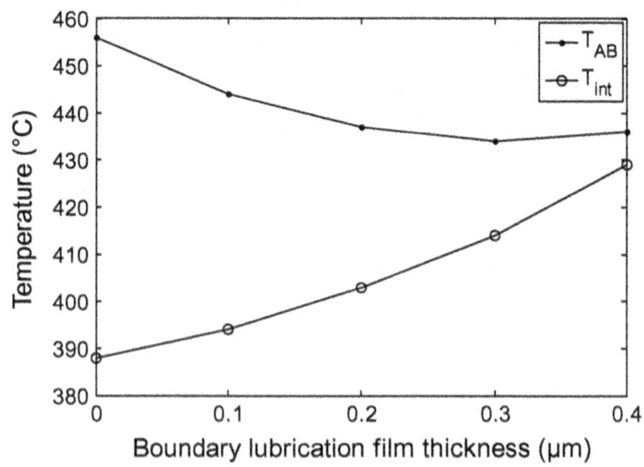

Fig. 6.2 The average temperature at shear plane and tool-chip interface under different boundary lubrication film thicknesses

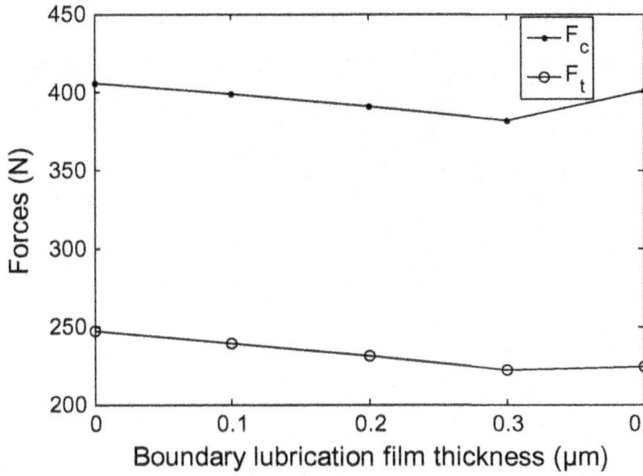

Fig. 6.3 Relationship between cutting forces and boundary lubrication film thicknesses

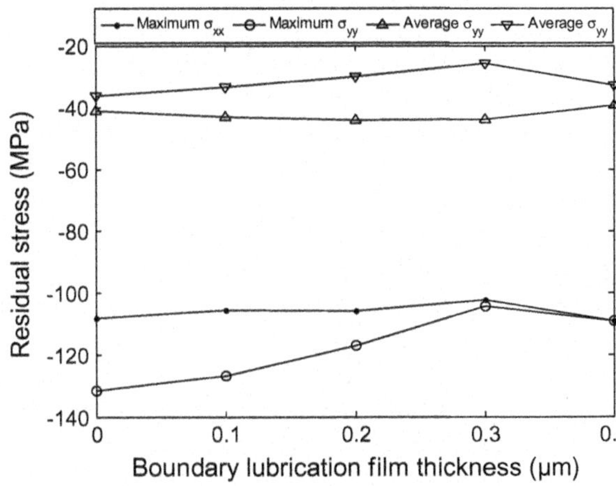

Fig. 6.4 Relationship between residual stresses and boundary lubrication film thicknesses

residual stress decreases with the increase of the boundary lubrication film thickness. When it increases to a certain extent, the maximum radial compressive residual stress increases accordingly. This change trend is consistent with the change trend of cutting force, it can be seen that the maximum compressive residual stress is mainly determined by the mechanical load. The change trend of the tangential maximum residual stress is not obvious. From Fig. 6.4, it is also found that the effect of the boundary lubrication film thickness on the average residual stress is not obvious.

6.2.2 Influence of Air-Oil Mixture Ratio

From the analysis in Sect. 2.3.2, it is known that the air-oil mixture ratio mainly affects the heat transfer coefficient in the temperature model, thus affecting the thermal stress in the cutting process. When the air-oil mixture ratio decreases, that is, the amount of air injected into the cutting area per unit time increases, and the heat transfer rate increases. Therefore, this section will discuss the effect of the air-oil mixture ratio on the cutting performance in MQL machining. Five different levels of air-oil mixture ratios are selected. The cutting speed is 1.5 m/s, the feed rate is 0.1 mm/r, the width of cut is 2 mm, and the flow rate of the cutting fluid is 32 ml/h. The corresponding prediction results are shown in Table 6.2. The maximum residual stress refers to the maximum residual compressive stress. The average residual stress refers to the average value of the residual stress from the subsurface layer to the depth of 100 μm. From Table 6.2, it can be seen that air-oil mixture ratio has no effect on the friction coefficient based on the prediction model.

The relationship between the heat transfer coefficient and the average temperature at shear plane and tool-chip interface is shown in Fig. 6.5. It is found that both the average temperature at the shear plane and at the tool-chip interface decreases with the increase of the heat transfer coefficient. It can be inferred that the heat transfer coefficient has a significant effect on the cooling in MQL machining, and its effect on the temperature at tool-chip interface is more significant than its effect on the temperature at shear plane.

Figure 6.6 shows the relationship between the heat transfer coefficient and cutting forces. It is found that both the tangential force and the axial force have a slight increase with the increase of the heat transfer coefficient. This is because when the heat transfer coefficient increases, the cooling effect increases, and the temperature decreases. The hardness and stiffness of the material increase, which will lead to the increase of the cutting force. From the analysis above, it can be concluded that the air-oil mixture ratio mainly affects the cooling effect of MQL.

Figure 6.7 shows the relationship between the heat transfer coefficient and residual stress. It can be seen that with the increase of the heat transfer coefficient, the maximum compressive residual stress increases first, and then decrease. While there is no obvious trend of the average residual stress with the heat transfer coefficient change. From Fig. 6.7, it can be seen that in order to obtain a larger compressive residual stress, there is a certain optimization range for the heat transfer coefficient. It is also found that the heat transfer coefficient corresponding to the maximum compressive residual stress is different from that corresponding to the average residual stress. Therefore, the different optimization objectives of residual stress determine the different selection of the parameters.

From the above discussion, it is known that when a smaller flow rate of MQL and a moderate air-oil mixture ratio are selected, a larger compressive residual stress can be obtained.

6.3 Influence of Cutting Parameters

Take the lathe turning of AISI 4130 alloy steel with non-coated carbide tool as an example, the workpiece is with an outer diameter of 63.5 mm, a tube wall thickness of

Table 6.2 Detailed predicted results under different air-oil mixture ratios

Air-oil mixture ratio level	H_{eff} (J/m^2 S °C)	μ	T_{AB} (°C)	T_{int} (°C)	F_c (N)	F_t (N)	Maximum value (MPa)		Average value (MPa)	
							σ_{xx}	σ_{yy}	σ_{xx}	σ_{yy}
1	470613.8	0.5753	495	486	368	217	−70.9394	−48.1608	−31.5718	−10.9555
2	627485.0	0.5753	466	444	380	224	−78.9139	−80.6867	−38.0449	−19.9493
3	784356.3	0.5753	437	402	391	230	−105.6719	−116.7954	−44.0927	−44.0927
4	941227.6	0.5753	408	361	403	237	−132.0838	−157.9656	−43.4378	−38.3999
5	1098099	0.5753	379	319	414	244	−94.2311	−113.2977	−34.6146	−35.5794

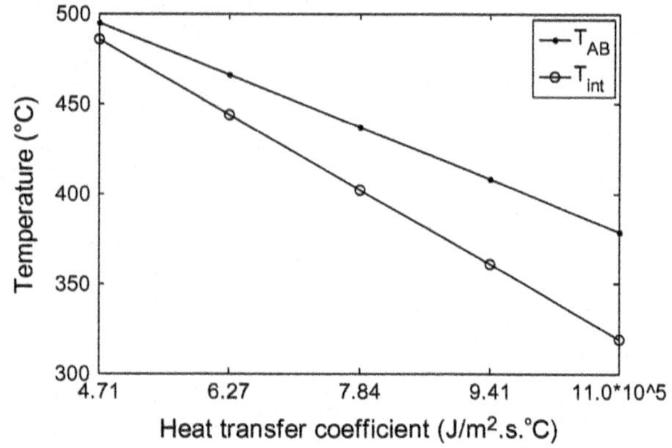

Fig. 6.5 Relationship between the heat transfer coefficient and average temperature at shear plane and tool-chip interface

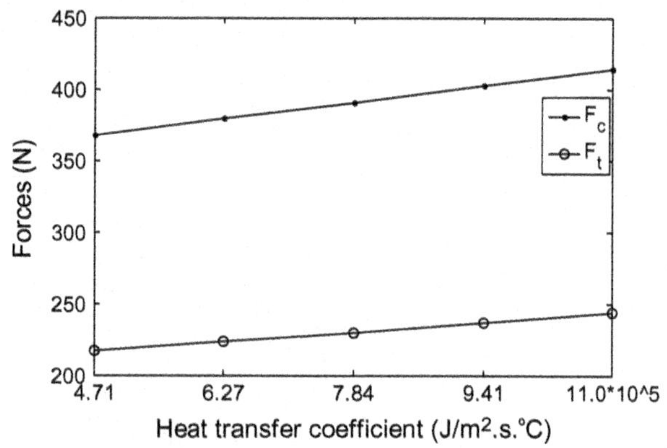

Fig. 6.6 Relationship between the heat transfer coefficient and cutting forces

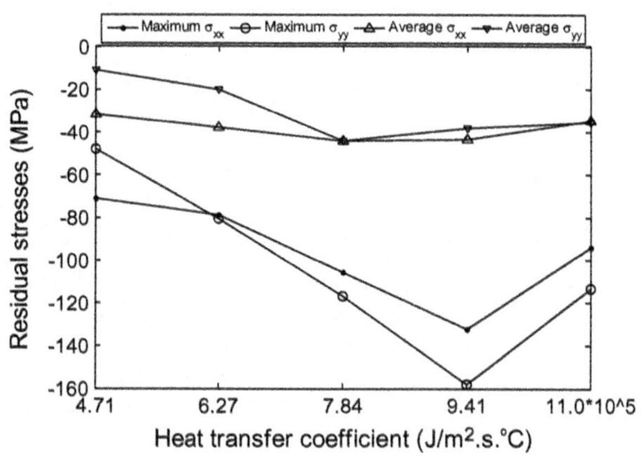

Fig. 6.7 Relationship between the heat transfer coefficient and residual stresses

4.775 mm, and a length of 60 mm, and its material properties are shown in Tables 5.3 and 5.4. The tool geometry parameters are shown in Table 5.5. The flow rate of MQL is 16 ml/h, and air-oil mixture ratio and air pressure are at moderate level. The effects of cutting speed, feed rate and the width of cut on the residual stress are discussed in details.

6.3.1 Influence of Cutting Speed

When the feed rate is 0.1 mm/r and the cutting width is 2 mm, the detailed prediction results under different cutting speeds are shown in Table 6.3. It can be seen the friction coefficient does not change obviously with the cutting speed.

Figure 6.8 shows the relationship between the cutting speed and average temperature at shear plane and tool-chip interface. It is found that both the average temperature at the shear plane and at the tool-chip interface increase with the increase of the cutting speed. This is because when the cutting speed increases, the amount of metal removal per unit time increases. Meantime, when the cutting speed increases, the heat generated during the cutting process cannot be dissipated in time, which will result in an increase in the temperature of the cutting area. From Fig. 6.8, it is also found that the average temperature at the shear plane is higher than that at the tool-chip interface, so it can be inferred that the heat in the cutting process mainly comes from the shear deformation, and the heat generated by the shear deformation is greater than that generated by the friction between the tool and the chip.

Figure 6.9 shows the relationship between the cutting speed and cutting forces. It is found that the cutting force decreases with the increase of cutting speed. This is because when the cutting speed increases, the cutting heat generated per unit time increases, which resulting in a decrease in the hardness and strength of the workpiece material, and the cutting force decreases as well. From the analysis in Chap. 5, it is found that the changing trend of the cutting force with the cutting speed remains the same whether it is under dry, MQL or traditional flood cooling condition. Therefore, the changing trend of cutting force with the cutting speed has no connection with the lubrication condition.

Figure 6.10 shows the relationship between the cutting speed and residual stresses. It is found that both the maximum compressive residual stress and the average residual stress decrease with the increase of the cutting speed. This is because when the cutting speed increases, the cutting force decreases, which resulting in a general reduction in residual stress. Combined with the analysis of cutting speed, cutting force and cutting temperature, it can be seen that the maximum compressive residual stress and the average residual stress are mainly determined by the mechanical load.

Table 6.3 Detailed predicted results under different cutting speeds

V (m/s)	μ	T_{AB} (°C)	T_{int} (°C)	F_c (N)	F_t (N)	Maximum value (MPa)		Average value (MPa)	
						σ_{xx}	σ_{yy}	σ_{xx}	σ_{yy}
1	0.6720	388	271	444	291	−123.5783	−129.4021	−54.3139	−47.9075
1.5	0.5753	437	403	391	231	−105.6719	−116.7954	−44.0927	−29.9024
2	0.5752	519	485	359	212	−63.3318	−66.0580	−28.6843	−16.8119
2.5	0.5752	601	534	356	205	−24.8646	−55.8099	−15.1627	−16.7311
3	0.5751	653	607	306	181	−21.0719	−23.3882	−6.77137	−2.1662

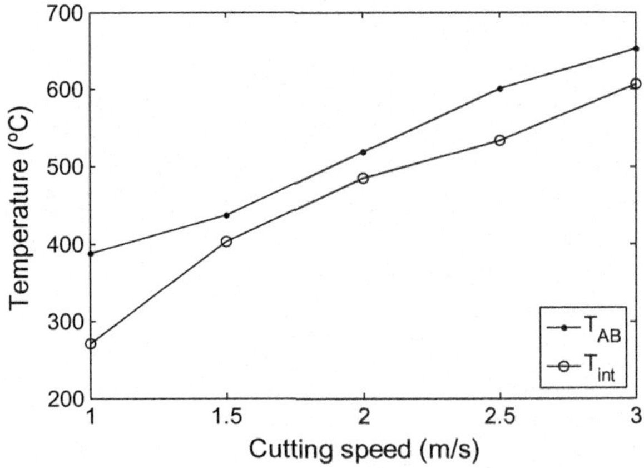

Fig. 6.8 Relationship between the cutting speed and average temperature at shear plane and tool-chip interface

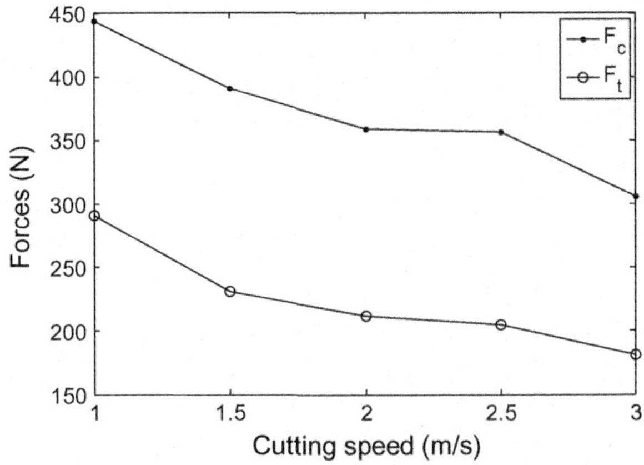

Fig. 6.9 Relationship between the cutting speed and cutting forces

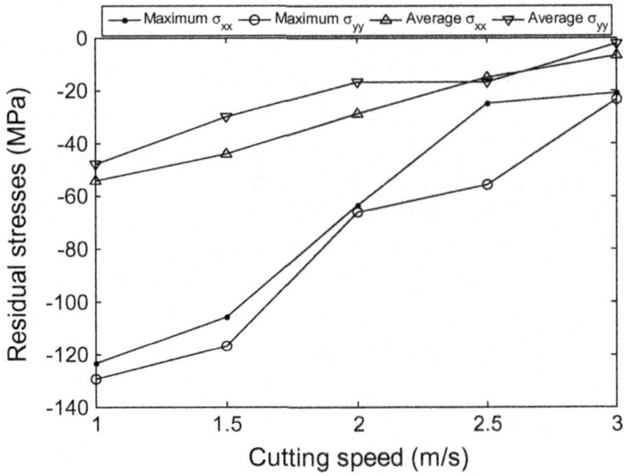

Fig. 6.10 Relationship between the cutting speed and residual stresses

6.3.2 *Influence of Feed Rate*

When the cutting speed is 1.5 m/s and the cutting width is 2 mm, the prediction results under different feed rates are shown in Table 6.4. It can be seen that the changing trend of the friction coefficient with the feed rate is not obvious. The cutting force increases with the increase of the feed rate, and the average temperature at the shear plane and at the tool-chip interface increases with the increase of the feed rate. This is because with the increase of the feed rate, the metal cutting amount per unit time increases, the cutting force increases, and the cutting temperature increases.

Figure 6.11 shows the relationship between the feed rate and the residual stresses. It is found that with the increase of the feed rate, the maximum compressive residual stress both in radial direction and in cutting direction have the same changing trend, first decrease and then increase. However, the changing trend of the average residual stress is not obvious with the increase of the feed rate. From Table 6.4, it is found that both cutting temperature and cutting force increase with the increase of the feed rate, and the increasing speed of cutting force is greater than that of cutting temperature. From the analysis results of residual stress, it can be known that the maximum compressive residual stress is mainly determined by the mechanical load.

6.3.3 *Influence of Cutting Width*

When the cutting speed is 1.5 m/s and the feed rate is 0.1 mm/r, the prediction results under different cutting widths are shown in Table 6.5. It is found that the cutting width has no significant effect on the friction coefficient. The average temperature at the shear plane and at the tool-chip interface decreases with the increase of the cutting width, but the trend is not very obvious. This is because when the cutting width increases, the cutting area and the heat dissipation area also increase accordingly, so the effect of the cutting width on the cutting temperature is not significant. The cutting force increases with the increase of the cutting width, this is because the cutting area increases with the increase of the cutting width, and the plastic deformation during the cutting process increases, which leads to an increase in the cutting force.

Figure 6.12 shows the relationship between the cutting width and the residual stresses. It is found that with the increase of the cutting width, the maximum compressive residual stress both in tangential direction and radial direction increase with the increase of the cutting width, while the average residual stress does not change significantly with the cutting width. On one hand, when the cutting width increases, the cutting force increases, and the stress generated by the mechanical load on the workpiece surface increases. On the other hand, since the average cutting temperature decreases when the cutting width increases, the stress on the workpiece surface caused by the thermal load decreases. Because of these two reasons, the effect of the width of cut on the average residual stress is not significant.

Table 6.4 Detailed predicted results under different cutting widths

Feed rate (mm/r)	μ	T_{AB} (°C)	T_{int} (°C)	F_c (N)	F_t (N)	Maximum value (MPa)		Average value (MPa)	
						σ_{xx}	σ_{yy}	σ_{xx}	σ_{yy}
0.1	0.5753	437	40 3	391	23 1	−105.6719	−116.7954	−44.0927	−29.9024
0.15	0.5752	54 1	482	510	274	−88.8287	−97.7258	−43.7486	−32.4547
0.2	0.5751	62 7	561	66 3	333	−51.0863	−69.4948	−36.0582	−40.3693
0.25	0.4835	616	685	700	294	−121.9177	−109.3953	−54.2298	−39.9699

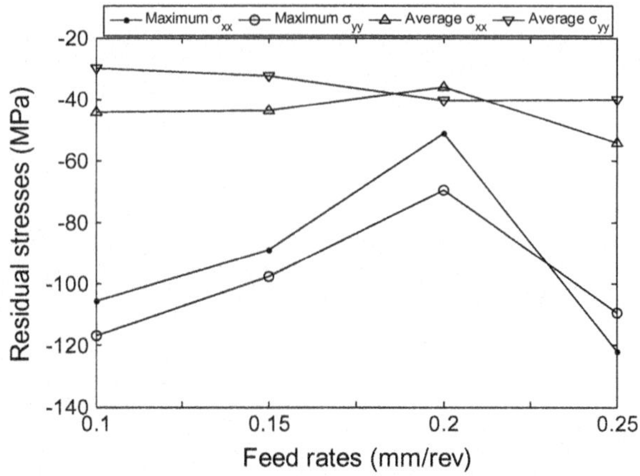

Fig. 6.11 Relationship between the feed rate and residual stresses

From the above discussion, it can be inferred that when the other cutting conditions are invariant, a larger compressive residual stress can be obtained by low cutting speed and small feed rate.

6.4 Influence of Tool Parameters

Take the cutting of AISI 4130 alloy steel with uncoated carbide tool as an example, the material parameters refer to Tables 5.3 and 5.4. The flow rate of MQL is 16 ml/h, air-oil mixture ratio and air pressure are kept at medium level, the cutting speed is 1.5 m/s, the feed rate is 0.1 mm/r, and the width of cut is 2 mm. The effects of tool rake angle and tool edge radius on the cutting force, cutting temperature, and residual stress are discussed in details.

6.4.1 Influence of Tool Rake Angle

When the tool edge radius is 30 μm, the prediction results under different rake angles are shown in Table 6.6. It can be seen that when the tool has a negative rake angle, the friction coefficient is small, and as the tool rake angle increases, the friction coefficient also increases. The average temperature at the shear plane and at the tool-chip interface does not change significantly with the tool rake angle. The cutting force decreases with the increases of the rake angle. This is because when the rake angle is negative, the extrusion deformation coefficient of the workpiece is large, so the cutting force is large as well.

Table 6.5 Detailed predicted results under different feed rates

w (mm)	μ	T_{AB} (°C)	T_{int} (°C)	F_c (N)	F_t (N)	Maximum value (MPa)		Average value (MPa)	
						σ_{xx}	σ_{yy}	σ_{xx}	σ_{yy}
1	0.5753	450	421	193	114	−93.3614	−100.0762	−41.454	−24.9292
2	0.5753	437	403	391	231	−105.6719	−116.7954	−44.0927	−29.9024
3	0.5753	427	388	592	350	−113.8923	−128.6733	−41.9967	−32.1156
4	0.5753	420	378	796	470	−120.6526	−138.7733	−42.6445	−34.296

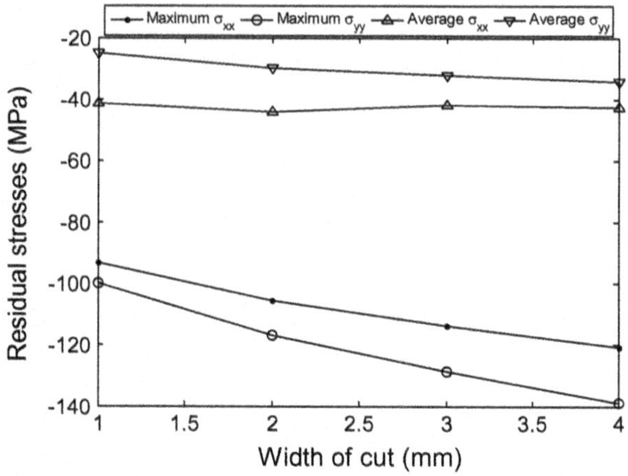

Fig. 6.12 Relationship between the width of cut and residual stresses

Figure 6.13 shows the relationship between the tool rake angle and the residual stresses. It is found that when the tool rake angle is negative, the maximum compressive residual stress is relatively large. When the tool rake angle is 0° and 5°, the changing trend of the maximum residual stress is not obvious. This is because when the tool rake angle is negative, the extrusion deformation of the material is serious, so the cutting force increases, and the stress generated by the mechanical load increases, which resulting in an increase of the compressive residual stress. The average residual stress does not change significantly with the tool rake angle. It can be inferred that a large compressive residual stress can be achieved by the large negative tool rake angle.

6.4.2 Influence of Tool Edge Radius

When the tool rake angle is 7°, the predicted results under different tool edge radiuses are shown in Table 6.7. It is found that the friction coefficient, the average temperature at the shear plane and at the tool-chip interface and the chip formation force are not affected by the tool edge radius. The plowing force increases with the increase of the tool edge radius. In this analysis, the total cutting force mainly includes the chip formation force and the plowing force, so the total cutting force increases with the increase of the tool edge radius.

Figure 6.14 shows the relationship between the tool edge radius and the residual stresses. It can be seen that the maximum compressive residual stress both in tangential and the radial direction increase first and then decrease with the increases of the tool edge radius. This is because when the tool edge radius increases, the resulting plowing force increases, and the stress on the workpiece surface due to the mechanical load increases, so the maximum compressive residual stress increases. The variation of the average residual stress with the tool edge radius is not very obvious. Therefore,

Table 6.6 Detailed predicted results under different rake angles

α (deg)	μ	T_{AB} (°C)	T_{int} (°C)	F_c (N)	F_t (N)	Maximum value (MPa)		Average value (MPa)	
						σ_{xx}	σ_{yy}	σ_{xx}	σ_{yy}
−5	0.4911	481	426	451	320	−160.7648	−175.4129	−28.3205	−45.4342
0	0.5123	452	406	423	272	−100.9171	−101.4141	−28.8785	−35.4119
5	0.6238	505	391	396	258	−89.1424	−107.5668	−36.1513	−27.4645

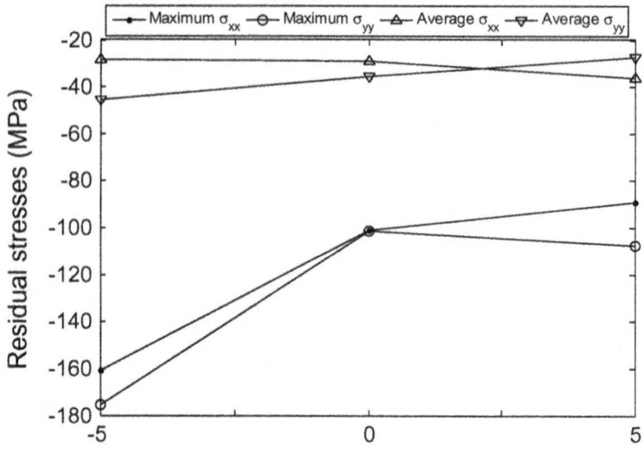

Fig. 6.13 Relationship between the rake angle and residual stresses

it can be inferred that a larger compressive residual stress can be obtained by large tool edge radius.

From the discussion above, within the explored experimental range, large compressive residual stress can be achieved by selecting small flow rate of MQL, medium air-oil mixture ratio, low cutting speed, small feed rate, and negative tool rake angle. Since the processing cost is considered in actual industrial production, decreasing the cutting speed and reducing the feed fate will lead to an increase in the processing cost. Therefore, larger compressive residual stress can be obtained by controlling the appropriate MQL parameters, such as low flow rate of cutting fluid.

6.5 Summary

Based on the verified residual stress prediction model, a sensitivity analysis of the residual stress in MQL machining is carried out. The effects of the MQL parameters, cutting parameters, and tool parameters on the residual stress of the machined surface are discussed. It is summarized as following:

(1) The maximum compressive residual stress first decreases and then increases with the increases of the boundary lubrication film thickness, but the depth of the maximum compressive residual stress does not change. The average residual stress in tangential direction first increases and then decreases with the increases of the boundary lubrication film thickness. On the contrary, the average residual stress in radial direction first decreases and then increases with the increases of the boundary lubrication film thickness. For different optimization objectives of the residual stress, the corresponding boundary lubrication film thickness is different.

(2) The maximum compressive residual stress first increases and then decreases with the increase of the air-oil mixture ratio, and the changing trend of the average residual stress is not obvious. Therefore, in order to obtain the large

Table 6.7 Detailed predicted results under different tool edge radius

r_e (μ m)	μ	T_{AB} (°C)	T_{int} (°C)	F_{chip_c} (N)	F_{chip_t} (N)	P_{cut} (N)	P_{Thrust} (N)	Maximum value (MPa)		Average value (MPa)	
								σ_{xx}	σ_{yy}	σ_{xx}	σ_{yy}
1	0.5753	437	403	358	151	1	3	−102.4268	−101.9236	−43.1501	−32.647
10	0.5753	437	403	358	151	11	27	−105.8799	−95.2711	−45.4876	−34.1364
20	0.5753	437	403	358	151	22	53	−118.1384	−116.2497	−43.7644	−34.6812
30	0.5753	437	403	358	151	34	80	−105.6719	−116.7954	−44.0927	−29.9024
40	0.5753	437	403	358	151	45	106	−91.1977	−110.1395	−40.3532	−21.2706

Fig. 6.14 Relationship between the tool edge radius and residual stresses

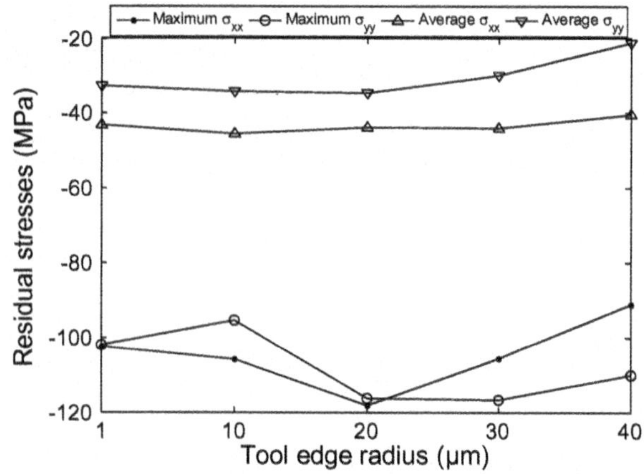

compressive residual stress, there is a certain optimal range for the air-oil mixture ratio.

(3) The maximum compressive residual stress decreases with the increase of cutting speed. But with the increase of feed rate, it first increases and then decreases. The maximum compressive residual stress is proportional to the cutting width. Therefore, larger compressive residual stress can be achieved by the selection of appropriate cutting parameters. The average residual stress does not change significantly with the cutting parameters.

(4) Large compressive residual stress can be obtained with negative tool rake angle. The maximum compressive residual stress first increases and then decreases with the increase of the tool edge radius, while the average residual stress does not change significantly with the tool geometry parameters.

(5) In the actual machining process, the machined residual stress can be controlled by adjusting the MQL parameters. Large compressive residual stress can be obtained with small flow rate of MQL and moderate air-oil mixture ratio.

The manufacturer's authorised representative in the EU is Springer Nature Customer Service Centre GmbH, Europaplatz 3, 69115 Heidelberg, Germany. If you have any concerns regarding our products, please contact ProductSafety@springernature.com

Printed and bound by CPI Group (UK) Ltd, Croydon, CR0 4YY
30/09/2026
02225308-0003